Education in the 80's:

MATHEMATICS

The Advisory Panel

Douglas B. Aichele, Head, Department of Curriculum and Instruction, Oklahoma State University, Stillwater

F. Joe Crosswhite, Professor, The Ohio State University, Columbus

Mary E. Froustet, Assistant Professor of Mathematics, Caldwell College, Caldwell, New Jersey

Gayle W. Harbo, Mathematics Department Chair, Austin E. Lathrop High School, Fairbanks, Alaska

Vernon R. Hood, mathematics instructor, Portland Community College, Oregon

Rose Ann Kafer, mathematics instructor, Olympia High School, Stanford, Illinois

Henry S. Kepner, Jr., Associate Professor, University of Wisconsin–Milwaukee

Alice W. Lahtela, former mathematics teacher and Mathematics Laboratory Coordinator, Capistrano Valley High School, San Juan Capistrano, California

Kil S. Lee, Assistant Professor of Mathematics Education, University of New Orleans, Louisiana

Edna McClung, mathematics teacher, Deming High School, New Mexico

Melfried Olson, Associate Professor of Mathematics and Mathematics Education, University of Wyoming, Laramie

Jesse A. Rudnick, Professor, Mathematics Education, Temple University, Philadelphia

M. Stoessel Wahl, Associate Professor of Mathematics, Western Connecticut State College, Danbury

Sue Pool White, mathematics teacher, Jefferson Junior High School, Washington, D.C.

James W. Wilson, Head, Mathematics Education Department, University of Georgia, Athens

June J. M. Yamashita, mathematics teacher, Kailua High School, Hawaii

Education in the 80's:

MATHEMATICS

Shirley Hill
Editor
University of Missouri-Kansas City

Classroom Teacher Consultant
Joseph F. Aieta
Mathematics Teacher
Weston High School
Massachusetts

National Education Association
Washington, D.C.

Copyright © 1982

National Education Association of the United States

Stock No. 3155-5-00 (paper)
3156-3-00 (cloth)

Note

The opinions expressed in this publication should not be construed as representing the policy or position of the National Education Association. Materials published as part of the NEA Education in the 80's series are intended to be discussion documents for teachers who are concerned with specialized interests of the profession.

Library of Congress Cataloging in Publication Data
Main entry under title:

Education in the 80's—mathematics.

  (Education in the 80's)
  Includes bibliographies.
  Contents: Mathematics education at the start of the decade: Status and prospects/James T. Fey—An interpretation of the results of the second NAEP mathematics assessment/Thomas P. Carpenter . . . [et al.]—Some basics: Computation, yesterday, today, and tomorrow/Marilyn N. Suydam.—[etc.]
  1. Mathematics—Study and teaching—United States—Addresses, essays, lectures. I. Hill, Shirley A. II. Title: Education in the eighties—mathematics. III. Series.
QA12.E38 510'.7'1 81-22487
ISBN 0-8106-3156-3 AACR2
ISBN 0-8106-3155-5 (pbk.)

Contents

Foreword   Joseph F. Aieta 7
Introduction   Shirley Hill 9

PART I—MATHEMATICS EDUCATION AT THE START OF THE DECADE

ONE — Status and Prospects
  James T. Fey 15

TWO — An Interpretation of the Results of the Second NAEP Mathematics Assessment
  Thomas P. Carpenter, Mary Kay Corbitt, Henry S. Kepner, Jr., Mary Montgomery Lindquist, and Robert E. Reys 24

PART II—SOME BASICS

THREE — Computation: Yesterday, Today, and Tomorrow
  Marilyn N. Suydam 36

FOUR — Measurement Is Basic
  Gwen Shufelt 46

FIVE — Computational Estimation Is a Basic Skill
  Robert E. Reys and Barbara J. Bestgen 53

SIX — Finding and Using Data: A Basic Skill
  Albert P. Shulte 65

SEVEN — Problem Solving: Searching for Solutions
  Mary Grace Kantowski 73

PART III—THE TOOLS OF TECHNOLOGY

EIGHT — Calculators in Schools: Thoughts and Suggestions
  Sherilyn Seitz and Terry E. Parks 81

NINE — Computers in School Mathematics
  J. D. Gawronski 88

PART IV—MATHEMATICS: THE CRITICAL FILTER

TEN — Women and Mathematics: Is There a Problem?
  Mary Schatz Koehler and Elizabeth Fennema 94

ELEVEN — The Case for a New High School Mathematics Curriculum
  Shirley Hill 108

The Contributors 118

Editor

Shirley Hill is Professor of Education and Mathematics at the University of Missouri–Kansas City. She is a coauthor of *Overview and Analysis of School Mathematics, K-12* and the author of books on geometry and logic.

Classroom Teacher Consultant

Joseph F. Aieta is a mathematics teacher in the Weston High School, Massachusetts.

Foreword

Everyone interested in mathematics education will find important perspectives for consideration in *Education in the 80's: Mathematics*. Whether the reader is more inclined to preserve the status quo or to call for major curriculum changes, this book will help provide a focus for some vital questions and related issues that are facing us in the 80's.

Public perceptions are that the quality of education is slipping, and school critics continue to point to declining test scores. Whether or not these test scores are true indicators of quality, teachers have felt pressure to prevent further loss of public support. Some of the conservatism reflected in the "back-to-basics" movement can be traced to teachers' perceptions about public expectations of the schools. But has the renewed attention to basic skills been an unqualified success? What do formal assessments and informal surveys tell us about student achievement and attitudes? By adopting a narrow definition of basic skills, do educators pay a price? What relative emphasis should be given to computation, estimation, measurement, problem solving, and data analysis? If each of these basic areas is to be developed and allowed to flourish in our programs, then some shifts in priorities must be considered by educators and publishers.

As we entered the 80's we also heard claims that our education programs have not been keeping up with the country's present and future needs for trained personnel in technical and technological fields, particularly in engineering and computer science. Projections of continued shortages of personnel throughout the decade have led some commentators on the international scene to worry that we are in danger of losing our technological edge. Has lack of student motivation, or our failure to require more mathematics and more science, contributed to this situation? Why does the "critical filter" of course work in secondary mathematics screen out so many more women than men from a significant number of college majors and subsequent careers? How can the schools intervene to encourage young adolescent women to pursue mathematics in the upper grades of high school?

It appears certain that calculators and computers will become more and more prevalent on the job and in the home. Educators and employers are currently promoting computer literacy among our citizenry. Will the integration of computers into the curriculum improve student problem-solving skills? Can computers help teachers improve student motivation? Will the extensive use of calculators and computers retard computational skill development? What types of computer applications in the schools make the most sense? To what extent should calculators and computers be regarded as objects of study in their own right? As a matter of fact, many college professors and secondary school teachers are more concerned about students who lack proficiency in algebraic manipulations than they are about students who enter college lacking computer literacy. At the entering level, calculus continues to dominate college mathematics; it also exerts the strongest influence on the last two years of the high school curriculum. If students return to their high schools reporting that they were well prepared for calculus, will teachers and administrators risk changing a curriculum that appears to be appropriate for those students? How can the promotion of probability, statistics, and computer science make a significant impression on curriculum decisionmakers? What should the schools provide for those students who are not planning to attend college?

This compendium of questions represents some of the issues addressed in this book. Several articles provide in-depth discussions of one major topic; others contain insights that cut across a number of separate issues. The authors seek to inform the reader. They propose no simplistic solutions. Veterans of curriculum campaigns in the 60's and 70's know that the words "crisis" or "revolution" do not motivate many people to seriously reconsider curriculum, teaching methods, or aids to teaching. In contrast to the post-Sputnik days of more robust support systems for teachers and schools, it is difficult today for mathematics teachers to maintain contact with researchers, innovators, and sources of inspiration in the field. *Education in the 80's: Mathematics* is one valuable way for educators to keep abreast of the work of authors who are truly concerned with the future of mathematics education.

Joseph F. Aieta
Mathematics Teacher
Weston High School
Massachusetts

Introduction

The decade of the eighties promises to be a period of excitement and ferment in school mathematics. There is a growing recognition among the public and among employers of the expanding need for mathematical skill and expertise, and of the importance of general mathematical and computer literacy to the consumer and citizen.

Support for the perceived importance of mathematics to individuals and to society includes recent Gallup polls (at 97 percent, mathematics ranked highest among subjects deemed essential for all high school students) and a major report, *Science and Engineering Education for the 1980's and Beyond,* prepared by the National Science Foundation and the Department of Education at the request of President Carter.

But the increasing awareness of growing societal demand for mathematical knowledge and skill is coupled with the widely held perception that our mathematics education is falling short. However valid this perception may or may not be, a number of recent reports coming after a publicized decade-and-a-half decline in SAT scores have stirred up cries of "crisis." These reports have described the intensive and extensive mathematics study in the schools of countries such as the Soviet Union, East Germany, and Japan. The comparisons are primarily quantitative—our students spend a considerably smaller proportion of their schooling in mathematical study—but there is some evidence of qualitative differences. Content descriptions indicate that the primary and middle school student in the Soviet Union, for example, has a much broader exposure to mathematical topics than the student in the United States, where elementary mathematics instruction concentrates heavily on arithmetic.

The picture is extraordinarily complex; nonetheless we find a predictable reaction: the perceived state of affairs in school mathematics is linked to national concerns for productivity and humanpower needs and technological dominance, and the inference is that "there is a crisis."

Shades of Sputnik! At this point the reader may have a feeling of déjà vu, but as James Fey points out in his article in Part I of this book, the conditions are vastly different from those in 1957. The schools are in a financially difficult position, public support for education in general and for teachers in particular is not what it should be, and, in an era of budget cutting, the federal government is most unlikely to pick up the tab for crash programs to bolster mathematics and science education.

Nevertheless, there are factors and trends that will certainly dictate major changes in mathematics teaching and curricula. One of these is the revolutionary effect on society of technological advances, in particular the computer. Electronic marvels, whether the pocket calculator for routine calculations on the job or for everyday living, or high-powered computers which change the very nature of handling information and solving problems, are our civilization's dominant tools. Educational objectives must comprehend the uses and influences of these devices, as J. D. Gawronski discusses in Part III. Despite the lingering worries of many parents and some teachers that calculators will make computational cripples of our students, the schools cannot ignore the fact that most complex computation in real-life settings is performed by using calculators. In another article in Part III, Sherilyn Seitz and Terry E. Parks describe activities from their direct experience in which classroom use of calculators reinforces but does not replace the learning of computational skills.

A realistic appraisal of the role of calculators in everyday life and in classroom problem solving will bring into focus the necessity for increasing instructional emphasis on skills of estimation. Several of the authors in this book make this point; and in their article Robert E. Reys and Barbara J. Bestgen discuss research support and suggestions for teaching the skill of computational estimation.

Another trend that may have peaked but that still dominates public pressure on schools is the "back-to-basics" movement. Mathematics teachers finally seem to be getting the point across that the "basic" or "essential" in mathematics goes well beyond a mastery of computation. Several categories of skill or knowledge are basic in mathematics. And changes in technology make some former basics obsolete. Skills are tools. They are basic only as long as the times demand them.

Part II describes several areas that should be considered basic in mathematics study. Marilyn N. Suydam discusses the role of computation today and tomorrow, in a historical context. Gwen Shufelt shows the development of measurement concepts from early intuitive activities to more sophisticated applications in high school. As previously mentioned, the Reys and Bestgen article presents the case for estimation skill as essential to all students.

In this day and age, it is difficult to imagine a more important consumer skill than the ability to understand information presented statistically. Today's citizen needs to know how to find data, to organize data, to present data in a clear and sometimes graphic manner, and to draw correct inferences from data. In short, informal statistics is a basic skill as Albert P. Shulte proposes in his article.

In her article on computation, Marilyn N. Suydam provides evidence that problem solving has always been the stated goal of school mathematics, while skills are the means to this end. Evidence accumulates, however, to show that school mathematics has not been fully meeting this goal and that emphasis has shifted to the means, away from the end. Recently, there has been a renewed interest in giving attention to the application of mathematics to the solution of problems. Support is strong for instruction in those skills and higher-order cognitive processes that are brought to bear on problem solving. Mary Grace Kantowski discusses what must be the ultimate "basic"—problem solving.

Another apparent trend is the criticism or at least questioning of the adequacy of present high school requirements in mathematics. Recently, surveys, reports, and conferences (e.g., the Priorities in School Mathematics project, the PRIME–80 conference of the Mathematical Association of America, the National Science Foundation-Department of Education report, *Science and Engineering Education for the 1980's and Beyond*) provide evidence that for a growing number of people and groups, the typical one-year requirement in grades 9 to 12 appears insufficient for a future which promises increasing uses of mathematics in everyone's life. Are students of high school age making informed choices about mathematics electives?

Sociologist Lucy Sells has used the term "critical filter" to define the role of mathematics in providing options in further education and careers. This idea is central to an awareness of the injustice represented by the glaring underrepresentation of minorities and women in mathematics-user positions and careers and in advanced mathematics study. Mary Schatz Koehler and Elizabeth Fennema in Part IV ask, "Is there a problem concerning female participation in mathematics?" and then explore the factors basic to the underrepresentation of females in mathematical pursuits.

The last article in the book discusses the suggestion that more mathematics be required in high school for all students. It makes the point that if this happens, the program will need to be diversified to meet a wide variety of needs, interests, and ability levels.

Finally, change in mathematics education may be hastened by the accumulation during the past five or six years of information about mathematics instruction and its outcomes. In the first article, James T. Fey gives an account of several important studies carried out between 1975 and 1979, which he uses to present the status of mathematics education and its future prospects.

One report which has already had considerable influence is the presentation of the results of the second assessment in mathematics of the National Assessment of Educational Progress. The second article, written by a team funded by NSF to interpret the data—Thomas P. Carpenter, Mary Kay Corbitt, Henry S. Kepner, Jr., Mary Montgomery Lindquist, and Robert E. Reys—summarizes the implications of these results.

Two documents referred to in several articles should be mentioned here. In 1979 and 1980, the National Council of Teachers of Mathematics conducted an extensive survey of the priorities and preferences for mathematics instruction of several populations, both professional and lay. Funded by the National Science Foundation and entitled *Priorities in School Mathematics,* the study is widely known as the PRISM report. It provides an interesting view of the important similarities and differences in various groups' priorities. It also provides an important part of the data base for the second document, published in 1980 by the National Council of Teachers of Mathematics. This document, *An Agenda for Action: Recommendations for School Mathematics of the 1980's,* is the organization's agenda to deal with needed change and future directions for mathematics programs.

It may be instructive and interesting to consider the merits of these recommendations against the backdrop of ideas and issues presented by the authors of this book. Accordingly, a summary of the eight major categories of the Council's recommendations follows.

The National Council of Teachers of Mathematics recommends that—

1. Problem solving be the focus of school mathematics in the 1980's.

2. Basic skills in mathematics be defined to encompass more than computational facility.

3. Mathematics programs take full advantage of the power of calculators and computers at all grade levels.

4. Stringent standards of both effectiveness and efficiency be applied to the teaching of mathematics.

5. The success of mathematics programs and student learning be evaluated by a wider range of measures than conventional testing.

6. More mathematics study be required for all students and a flexible curriculum with a greater range of options be designed to accommodate the diverse needs of the student population.

7. Mathematics teachers demand of themselves and their colleagues a high level of professionalism.
8. Public support for mathematics instruction be raised to a level commensurate with the importance of mathematical understanding to individuals and society.

Shirley Hill

ONE

Status and Prospects
James T. Fey

For those who study the history of U.S. education, the status of school mathematics today must bear striking resemblance to the critical period just 25 years ago. In 1956, disappointment with the mathematical preparation of entering college students led to the formation of a College Board Commission on Mathematics. The challenge of scientific competition with the Soviet Union, sparked by the 1957 Sputnik crisis, led to urgent calls for strengthened and modernized school mathematics curricula. At the same time, U.S. schools faced a critical shortage of qualified mathematics teachers, and those already in schools needed extensive reeducation in content and pedagogy.

Today the challenges in school mathematics are remarkably similar. Reports of student achievement at all levels seem consistently discouraging; sovietologists have recently reported dramatic progress in the mathematics education achieved by that country; electronic technology promises to bring fundamental changes to curricula and teaching methods; and again there is a critical national shortage of qualified mathematics teachers. In 1980 the Secretary of Education and the Director of the National Science Foundation (NSF) submitted a report to President Carter highlighting these and other problems and calling for major government action in science education.

If one is inclined to reason by historical analogy, the task of planning school mathematics for the 1980's offers an attractive opportunity. It seems appropriate to ask: Which responses to the earlier crisis worked? Which failed? How are conditions today different from those of the late 50's, and what impact will these different conditions have on efforts to bring about positive change? Fortunately, in mathematics education, there are several recent surveys and reflective analyses that provide the kind of background understanding needed for thoughtful planning.

RECENT THEMES IN SCHOOL MATHEMATICS REFORM

In 1974 the Conference Board of the Mathematical Sciences organized a National Advisory Committee on Mathematics Education (NACOME) to prepare an *Overview and Analysis of School Mathematics K–12*. One of the first tasks of that committee was to survey the efforts and accomplishments of the "new math" era. The crisis in school mathematics of the late 1950's led to dozens of major curriculum development projects, thousands of institutes and conferences for teachers, major research and evaluation projects, and lively controversy about the wisdom of various reform efforts. For the first time in history the federal government played a major role in guidance and financial support of educational development, and school mathematics became a national political issue.

As NACOME pointed out, from 1955 to 1970 reform efforts in school mathematics were intended to produce high-quality curricula for college-capable students. The innovative programs were designed with major advice from university and industrial mathematicians and psychologists. The central themes included emphasis on student understanding of mathematical methods, use of unifying concepts and structures, increased precision of language and reasoning, acceleration of many traditional topics or courses, introduction of new topics (notably calculus and statistics), and deletion of outdated material. The course content and sequence changes were accompanied by recommendations favoring laboratory and discovery instruction.

Beginning about 1970 a new set of background conditions and pressures led school mathematics development in different directions. Attention shifted to programs for less able students, minimal competence for job entry, applications of mathematics, and school accountability through extensive achievement testing. From focus on advanced topics for college-bound students, energy and attention shifted to basic skills appropriate for all students. The new direction in curriculum gained a catch-phrase label, "back to basics," and an accompanying instructional theory of behavioral objectives and individually prescribed learning programs.

Like its predecessors in the educational limelight, the basic skills thrust was prompted by disappointing student achievement reports and subsequent public concern. The concern led to curriculum advisory conferences and development projects. Changes in school texts were soon noticed, and these changes provoked controversy about proper minimal competence goals and methods of assessing achievement. This pattern of evolving change follows that of the new math era a decade earlier. However, the back-to-basics movement reveals effects of different change agents. Professional mathematicians played little role in stimulating or guiding the initiatives. The specific characteristics of back-to-basics programs seem more responsive to the demands of a concerned public and of some classroom teachers.

NET EFFECTS OF REFORM—SCHOOL MATHEMATICS TODAY

As mathematics educators have debated the value of successive innovations—new math, discovery teaching, behavioral objectives, individualized instruction, basic skills programs—perceptive observers have frequently questioned the extent to which any of these proposed innovations became part of daily practice in normal school classrooms. As the 1975 National Advisory Committee commented, "Serious gaps in the available information prevent definitive resolution of many questions." (2, p. xiv). That committee worked hard to assemble bits and pieces of pertinent information and to indicate the kinds of data needed. Since its report the National Assessment of Educational Progress (NAEP) has released two studies of school mathematics achievement; the National Science Foundation has published two extensive surveys of curricula, teaching, and teacher characteristics; and the National Council of Teachers of Mathematics has completed an extensive survey of *Priorities in School Mathematics.* A synthesis of these findings gives valuable empirical evidence on the effects of recent change efforts, the status of school mathematics today, and the prospects for future change.

Curricula in Use

After assessing the curricular initiatives of the new math era, the NACOME report concluded that "From a 1975 perspective the principal thrust of change in school mathematics remains fundamentally sound, though actual impact has been modest relative to expectations." (2, p. 21). The Committee further conjectured that by 1975 a back-to-basics mood had begun to influence curricular decisions at all grade levels.

The NACOME conclusions about curriculum practice were based on study of textbooks, published syllabi, a National Center of Education Statistics survey of course offerings and enrollments, and informal reports from leaders in the field. The 1977–78 NSF surveys gave more extensive data on these same indicators, but left open for conjecture the actual substance of courses delivered to students. For example, elementary texts usually contain material on geometry and statistics, but many experts in the field suspect that this material is routinely skipped in favor of emphasis on arithmetic computation. At the secondary level algebra texts contain material on trigonometry and probability, but this content is also frequently omitted in favor of further practice with algebraic manipulations. Time and time again these conjectures were confirmed in reports to NSF case study observers with comments such as the following from elementary teachers:

> Modern mathematics? I dislike it. . . . [The text] shows three ways when one will do. The brass tacks are learning addition and subtraction. That's it.
>
> This book has too much esoteric garbage in it. It is simply too hard. The geometry is silly [to try and teach] even for our best third graders. So we all skip it.
>
> We are fortunate not to have gone way out for the new math. We have stuck to the basics throughout it all and the results that are coming show we were right. (6, pp. 31–33)

Secondary teachers applauded these trends in elementary mathematics. These sentiments therefore confirm NACOME conjectures that new math era changes have had only modest impact on the content of school mathematics and suggest an obvious explanation—the innovators failed to win the minds and hearts of classroom teachers.

Teaching and Teachers

Over the past 25 years teachers in grades K–12 have received recommendations for change in teaching methods that nearly equal the scope of curriculum proposals. But, despite urging to use discovery, laboratory activities, individualized programs, computers, and other promising approaches, the weight of evidence suggests that school mathematics teaching today follows a limited traditional pattern: first, answers to the previous day's homework are given, with representative problems worked by the teacher or an able student; then, teacher-directed explanation and questioning are used to present material for the next day's assignment; finally, students begin work on the assignment at their seats. One observer in the NSF studies noted that "Although it seemed boring to me, students and teachers seemed comfortable with it. Apparently it fulfills student expectations and provides the students opportunity for closure." (6, p. 6). The teachers supported their choice of teaching methods with comments like the following:

> We've found that traditional methods of instruction work. This is the way it was taught to us in high school and the way it was taught in college and the way it works for us. . . . I don't think kids can handle inquiry. (6, p. 11)

The classroom practices of teachers undoubtedly reflect their knowledge and beliefs about mathematics and how students learn. The NSF surveys indicate that, as of 1977, secondary mathematics teachers averaged 12 years of experience, half held a degree beyond the bachelor's degree, and about 40 percent were taking a course for college credit. A majority of mathematics classes were taught by men and by teachers for whom mathematics was their only subject area of responsibility. An overwhelming number of those teachers felt adequately qualified to carry out their teaching assignments.

When teachers in grades 7–12 were asked to specify areas in which they would like assistance, they mentioned learning new teaching methods, information on instructional materials, implementing discovery/inquiry methods, using manipulatives, working with small groups, and articulation across grade levels. At most, however, 42 percent of the teachers mentioned any one of these areas. When asked to rate the seriousness of various potential problems, these same teachers consistently stressed lack of materials for individualizing instruction, lack of student interest in the subject, inadequate student reading abilities, and too-large class size. Eighty percent reported low student interest to be a problem and 90 percent reported inadequate reading abilities to be a problem.

At the elementary level also, the average teacher has over 12 years of experience. In all likelihood the mathematical competence of these teachers is much greater than that of teachers of 20 years ago, and, for the most part, the teachers feel competent to do a good job in mathematics. However, state supervisors of mathematics saw lack of teacher interest in mathematics and inadequate preparation to teach mathematics as the most serious problems in grades K–6. This contrast suggests that K–6 teachers and their immediate supervisors believe in the computation curriculum and in "tell-and-drill" methods of instruction—a pattern of beliefs that will not please many mathematics educators, and one that constitutes a formidable barrier to change (7).

These survey data give a sketchy quantitative outline of mathematics teachers' backgrounds and concerns, but they only begin to tell the story of teachers' attitudes and beliefs that emerges from consistent findings of case study interviews. With near-perfect regularity, teachers at all grade levels support the trend toward traditional content, instructional methods, and higher standards of student performance. They believe that mastery of certain skills is an essential prerequisite for concept learning and creativity. While frustrated by inability to motivate students with genuine applications, most teachers find virtue in mathematics as an arena for teaching logical thought, problem solving, and careful hard work. Like teachers in every subject area, mathematics teachers talk of going stale, of losing enthusiasm for their task. This feeling is sometimes expressed in negative feelings toward students (they don't care or try, they're spoiled); toward school administrators, college and university training programs, and the community.

Despite the frequent complaints, the overall mood that filters out of the NSF status surveys is well captured by the following summary:

> They saw themselves in a serious, not very exciting business; the business of education. They saw themselves as pretty good businessmen, wishing that times would change for the better, but confident that they could deliver on their promises and pretty well satisfied that there is not really a better way to run the shop. (6, p. 25)

This picture of mathematics teachers today, combined with the earlier profiles of curriculum and instruction in mathematics, shows the realities with which any attempted change in mathematics education must deal.

Students and Their Achievement

Changes in curriculum content or teaching methods will prompt vigorous controversy among mathematicians and classroom teachers of mathematics, but the only certain cause of public debate concerning school mathematics is a report of student achievement test scores. For the past ten years this news seems to have been all bad. College aptitude test scores have declined steadily, failure rates on minimal competence tests have been distressingly high, and student performance at most grade levels has declined in relation to national norms of earlier periods. All these data add up to a pervasive public impression that school achievement is far below what it reasonably ought to be.

The actual profile of student achievement results is not so consistent as impressions suggest, nor are the explanations for declines in several areas simple. For instance, when NACOME studied available mathematics test data in 1975, the committee noted that declining performance was concentrated at the secondary grade levels, that performance on basic computational skills was not so disappointing as on more complex problem-solving tasks, and that there was little convincing evidence that new math curricula were major contributors to changing patterns of achievement. The recently reported second national assessment of mathematics revealed similar patterns—computational skills with whole numbers and decimals seemed satisfactory, but problem-solving performance was poor.

Several prestigious committees, each searching for causes of the trends, have analyzed the declining levels of performance on college entrance examinations. Each report has concluded that the causes are complex and interrelated. Since the declines in mathematics performance have been consistently less than those in verbal tests, one could reasonably argue that mathematics teaching has resisted a general decline in academic achievement.

When classroom teachers are asked to conjecture causes of declining student achievement, they place poor student motivation at the top of the list. While there is clear evidence that student attitudes toward mathematics decline throughout secondary school, it is not so clear that the situation is much different now from what it was in the past. Others suggest that fewer students are taking advanced mathematics in high school and that graduation or college entrance requirements in mathematics have been reduced over the past 10 years. Again, the evidence in favor of this explanation is not clear. The NSF surveys found 56 percent of all school districts requiring one mathematics course at most for high school graduation and this course could be something like general mathematics. Enrollment figures suggest, however, that most students are really taking much more mathematics.

At the elementary grades it is very clear that schools spend considerable time on mathematics, second only to language arts and far ahead of any other subject area. This probably represents a significant change from the period 20 years ago and may well account for the generally solid performance in grade school mathematics.

Social Context of Schooling

When specialists in mathematics education seek ways to improve curricula, teaching, and student achievement, they tend to look within the discipline of mathematics and the mathematics classroom for ideas. But trends such as teacher burnout or declining student achievement and motivation are not limited to school mathematics. They suggest that the effectiveness of any one special instructional program is affected by a complex of broad school and societal factors.

The new math era prompted by the Sputnik crisis of 1957 was sustained by broad public support. Governments invested large sums of money in curriculum development and teacher education; teacher salaries gained relative to other occupations; new and well-equipped schools were built at a rapid rate; parents encouraged their children to study as much mathematics as possible and to work hard in preparation for college entrance. When one looks at the societal support system for schools in general, and for mathematics education in particular, conditions today are distressingly different.

Education is no longer a growth industry; student populations are declining and resource allocations are tightly constrained. Schools have been asked to play leading roles in social integration, with the result that many teachers report chaotic school conditions not conducive to serious study. Instability in family life has also contributed to declining home interest in and support for schools.

The recent National Council of Teachers of Mathematics survey of *Priorities in School Mathematics* (PRISM) revealed that the public still places high value on mathematics education (5). However, the public view of curricular emphases tends to be very conservative and thus at odds with the priorities of leaders in the field.

PROSPECTS FOR THE 80'S

Nearly every survey of the past 25 years in mathematics education leads to the same simple conclusion: great ambitions, modest accomplishment. As we mentioned at the outset of this article, the field faces challenges today no less critical than those of that earlier period. Curricula and teaching methods must be totally reconstructed to take account and advantage of the new microelectronic world in which current and future students will live. The resources to carry out such fundamental reconstruction of school mathematics are not abundant, and the record of efforts during the new math and back-to-basics eras suggests that implementing change will not be easy. Despite the complaints of some that schools "experiment too much," the facts show that education is a very conservative institution. In the search for reasons why many teachers seem to have lost their enthusiasm and spirit of innovation, it has been suggested that a teacher's position is really very isolated. As the sole adult in a class of young people it is often difficult for teachers to maintain contact with the scientific and professional communities beyond the school. Some teachers have "kept a window on the larger world of ideas," but "most teachers have only a mirror that reflects the values and ideas already dominant in the public schools" (6, p. 18).

There are sharp differences between the priorities of the public, classroom teachers, administrators, and innovators in mathematics education. Furthermore, those in the classroom tend to dismiss outside opinion as impractical and out-of-touch; those outside tend to criticize teachers for stubbornly resisting any new ideas. This climate of distrust and criticism that bars cooperative work on important problems is a tragedy for mathematics education. Redevelopment of working relations among the various parties with interest in school mathematics must be high on the priority list of the profession.

The National Council of Teachers of Mathematics has set an Agenda for Action for the decade ahead. Perhaps with the insight of recent experiences and the combined energy of the many professionals who care about the health of school mathematics, it will be possible to look back at the end of the decade to a period of striking accomplishment.

REFERENCES

Information in this article is synthesized from a variety of original and interpretive sources; the principal papers are the following:

1. Carpenter, Thomas, et al. "Results of the Second NAEP Mathematics Assessment: Secondary School." *Mathematics Teacher* 73 (May 1980): 329–33.

2. Conference Board of the Mathematical Sciences National Advisory Committee on Mathematics Education (NACOME). *Overview and Analysis of School Mathematics K–12*. Washington, D.C.: Conference Board of the Mathematical Sciences, 1975.

3. Fey, James T. "Mathematics Teaching Today: Perspectives from Three National Surveys." *Mathematics Teacher* 72 (October, 1979): 490–504.

4. National Council of Teachers of Mathematics. *An Agenda for Action: Recommendations for School Mathematics of the 1980's*. Reston, Va.: The Council, 1980.

5. ———. *Priorities in School Mathematics: Executive Summary of the PRISM Project*. Reston, Va.: The Council, 1981.

6. Stake, Robert E., and Easley, Jack, eds. *Case Studies in Science Education*. Urbana, Ill.: University of Illinois, 1978.

7. Weiss, Iris. *Report of the 1977 National Survey of Science, Mathematics, and Social Studies Education*. Research Triangle Park, N.C.: Research Triangle Institute, 1978.

TWO

An Interpretation of the Results of the Second NAEP Mathematics Assessment*

Thomas P. Carpenter, Mary Kay Corbitt, Henry S. Kepner, Jr., Mary Montgomery Lindquist, and Robert E. Reys

Many questions are asked about student achievement and attitudes in mathematics. One of the primary sources that may be used to answer such questions is the mathematics assessment of the National Assessment of Educational Progress (NAEP). The results of the second mathematics assessment, conducted during 1977 and 1978, provide insight into what our students are and are not learning and also into their attitudes toward mathematics.

This article presents an interpretation of these results in the form of selected conclusions. In order to place the conclusions into perspective, it is important to understand the purposes of the assessment, the nature of the sample, and certain aspects of the assessment procedures.

One purpose of the assessment was to make available comprehensive data on educational attainments of young Americans. Thus, the exercises covered a wide range of objectives selected by panels of mathematicians, mathematics educators, classroom teachers, and lay people. Another purpose was to measure change. Some exercises were therefore designed to reflect changes in future curricula. For example, some decimal exercises were administered to 9-year-olds so that baseline data

---

*This article is based upon work supported by the National Science Foundation under Grant No. SED-7920086. Any opinions, findings, conclusions, or recommendations expressed here are those of the authors and do not necessarily reflect the views of the National Science Foundation.

would be available for comparisons at a later date. In other words, it was not expected that 9-year-olds would do well on decimal exercises, but that there would be a change on future assessments.

The sample, consisting of more than seventy thousand 9-, 13-, and 17-year-olds, was carefully selected to be representative of many facets of our population. The results provide an accurate sampling of U.S. elementary and secondary students rather than of special populations such as college-bound seniors. For example, the 17-year-old population consisted of students in the tenth to twelfth grades, about one-half of whom had had at least a half-year of general or business mathematics, about two-thirds of whom had had at least a half-year of algebra, but only about one-sixth of whom had had trigonometry.

Altogether, approximately 230 exercises were administered to 9-year-olds, 350 to 13-year-olds, and 450 to 17-year-olds. Since testing time was limited to 45 minutes for each participant, an item-sampling procedure was used to administer each exercise to approximately 2,400 respondents at each age level.

All exercises were administered by specially trained exercise administrators to groups of fewer than 25 students. To standardize procedures and to minimize reading difficulty, all exercises were presented on a paced audiotape as well as in exercise booklets. Both multiple choice and open-ended exercises were included. Scoring guides were developed for the open-ended exercises in order to identify the percentage of respondents making specific errors.

We have selected six conclusions as our interpretation of some of the results; other authors may reach other conclusions. These six conclusions, however, are generally supported by a wide range of exercises, not merely the illustrative exercises reported here.

SIX CONCLUSIONS

1. Students demonstrated a high level of mastery of addition, subtraction, and multiplication of whole numbers.

A great amount of instructional time is devoted to the skill of computing with whole numbers. The results of this assessment showed that this emphasis does produce students who can add, subtract, and multiply whole numbers.

The 9-year-olds had mastered the basic addition and subtraction facts such as $8 + 5$ or $13 - 6$, and the two older age groups had mastered all the basic number facts. About two-thirds of the 9-year-olds could perform simple addition and subtraction computation involving regrouping. For example, about 75 percent of them gave the correct response to 37 + 18. By age 13, almost all students could perform simple computations involving addition, subtraction, or multiplication. Most older students were successful with more difficult calculations such as those given in Table 1.

TABLE 1
Addition, Subtraction, and Multiplication Computation

Exercise	Percent Correct	
	Age 13	Age 17
4285 3273 + 5125	85	90
Subtract 237 from 504	73	84
671 × 402	66	77

The conclusions that follow indicate that the strength shown in these computations was not found in all other number areas. Performance was also lower on exercises assessing basic noncomputational skills. For example, only 4 percent of the 13-year-olds and 18 percent of the 17-year-olds found the area of a right triangle. In general, the only noncomputational skills for which students demonstrated a high level of mastery were those involving simple intuitive concepts or those skills they were likely to have encountered and practiced outside school. For example, 85 percent of the 13-year-olds could tell time and 81 percent of the 9-year-olds could read a bathroom scale.

While the results of the exercises involving addition, subtraction, or multiplication are positive, they must be considered in the light of the other results. If the amount of time spent on these skills prohibits the development of other basic skills, then it is necessary to examine not only priorities, but also the ways in which and the times at which these skills are developed.

2. Students experienced difficulty with division of whole numbers.

Both 13- and 17-year-olds had difficulty with division computation. Only about half of the students in these two age groups made the following calculation correctly:

$$28 \overline{)3052}.$$

About 30 percent of the 13-year-olds and 15 percent of the 17-year-olds missed even simple division exercises. With a calculator, however, over 80 percent of the 13-year-olds and over 90 percent of the 17-year-olds could do the more difficult division exercises. This raises some serious questions as to the productivity of time and effort spent drilling on division and whether that time and effort should be devoted to other topics. Clearly, the current approach to teaching division is not effective for most students.

The division algorithm, as well as most of the other algorithms taught in school, is designed to produce rapid, accurate calculation procedures. Given the widespread availability of hand calculators, it would seem that the continued emphasis on developing facility with computation algorithms should not be as high a priority as it was formerly. Certainly computation is important; but what is needed are algorithms that students will remember and will be able to generalize to new situations.

3. Students showed a lack of understanding of fractions, decimals, and percentages.

Students' performance showed a lack of understanding of basic concepts and processes associated with fractions, decimals, and percentages. This conclusion is drawn from examining computation, estimation, and word problems.

The results in Table 2 indicate that most students could add simple fractions with common denominators. However, when adding fractions with unlike denominators many used superficial manipulations. For example, 30 percent of the 13-year-olds and 15 percent of the 17-year-olds added the numerators and denominators to arrive at $\frac{2}{5}$ for the sum of $\frac{1}{2}$ and $\frac{1}{3}$. Notice that the complexity of the unlike denominators had relatively little effect on the students' performance. It appears that if students have learned and can recall an algorithm, they can successfully apply it. However, if they have not learned or cannot recall a mechanical algorithm, they cannot solve even simple problems that might be solved intuitively or by using simple models of fractions.

Contrast the results of the last exercise in Table 2 with the exercise in Table 3. This estimation exercise and the errors vividly point out the lack of understanding of fractions. More students could correctly compute the sum of fractions than could choose an estimate of the sum.

Further evidence of the lack of understanding is found when a fraction computation exercise is contrasted with a verbal problem. Given a simple verbal problem requiring fraction multiplication for its

TABLE 2
Fraction Addition Exercises

Exercise	Percent Correct	
	Age 13	Age 17
$\frac{4}{12} + \frac{3}{12} =$	74	90
$\begin{array}{r} 2\ 3/5 \\ +\ 4\ 4/5 \end{array}$	63	77
$\frac{1}{2} + \frac{1}{3} =$	33	66
$\begin{array}{r} \frac{7}{15} \\ +\ \frac{4}{9} \end{array}$	39	54

TABLE 3
Performance on a Fraction Estimation Exercise

ESTIMATE the answer to $\frac{12}{13} + \frac{7}{8}$. You will not have time to solve the problem using paper and pencil.

		Percent Responding	
		Age 13	Age 17
○	1	7	8
●	2	24	37
○	19	28	21
○	21	27	15
○	I don't know	14	18

solution, fewer than one-third of the 13- and 17-year-olds gave the correct response, despite the fact that approximately three-fourths of both groups could correctly multiply the fractions involved. These results indicate that students had no clear conception of the meaning of fraction multiplication, and therefore could not apply their skills to solve a simple problem.

If students have mastered basic decimal concepts, then operations with decimals are essentially the same as those with whole numbers. About half of the 13-year-olds and two-thirds of the 17-year-olds were successful on exercises assessing basic decimal concepts, and about the same number of students could also add, subtract, and multiply decimals. Performance on division exercises was much lower, however; the lack of understanding of both decimals and division could explain this level of performance.

Successful performance with percentages requires a firm foundation of fractions and decimals. Not surprisingly, then, performance on percentage exercises was extremely low. Table 4 gives illustrative exercises and results.

TABLE 4
OPERATIONS WITH PERCENTAGES

	PERCENT CORRECT	
	Age 13	Age 17
A. 30 is what percent of 60?	35	58
B. What is 4% of 75?	8	27
C. 12 is 15% of what number?	4	12
D. What is 125% of 40?	12	31
E. 6 is what percent of 120?	6	16

The importance of understanding may, in part, account for the difference in the level of performance between operations involving whole numbers and operations involving fractions, decimals, and percentages. Most assessment exercises indicated that students had learned the basic concepts underlying whole number computation, and had some notion of the place value concepts involved in the computation algorithms. As a consequence, performance on whole number computation exercises was, in our opinion, generally good. The results also suggested, however, that most students did not have a clear understanding of fractions, decimals, and percentages, and appeared to operate at a mechanical level. This lack of understanding resulted in relatively poor performance.

The same lack of understanding was also observed in other content areas such as measurement and probability, as well as in problem solving. Students perceived that understanding is an integral part of mathematics learning, as evidenced by the fact that 90 percent of the older age groups agreed with the statement, "Knowing why an answer is correct is as important as getting the correct answer." Their responses may reflect either their actual belief, or the fact that they had heard the statement and they considered agreement with it the "right" response. This second alternative gains credence when compared with the fact that about 90 percent of the two older age groups agreed that "There is always a rule to follow in solving mathematics problems." The students may be concentrating on mastering rules to the extent of ignoring concomitant understanding, because their experience dictates that right answers, usually obtained by means of rules, are rewarded.

4. Students were successful on one-step routine verbal problems, but showed a lack of basic problem-solving skills.

One of the consequences of learning mathematical skills at a rote, mechanical level is that students cannot apply the skills they have learned to solve problems. In general, NAEP results showed that the majority of students at all age levels had difficulty with any nonroutine problem that required some analysis or thinking. It appears that students have not learned basic problem-solving skills.

Problem solving is often equated with solving textbook verbal problems, but these were not the type of problems that caused difficulty. In fact, students generally were successful in solving routine one-step verbal problems such as those found in typical textbooks. Results suggested that if students understood the operation involved in routine one-step verbal problems, finding the solution presented no difficulty. For example, 38 percent of the 9-year-olds and 82 percent of the 13-year-olds correctly solved a problem similar to the following:

> Sue had 342 stamps in her collection. If 278 of them were U.S. stamps, how many were foreign stamps?

For the corresponding computation problem (342 − 278), 50 percent of the 9-year-olds and 85 percent of the 13-year-olds calculated the correct answer.

Although students could solve most simple one-step problems, they had a great deal of difficulty analyzing nonroutine or multistep problems. In fact, for a problem that required several steps or that contained extraneous information, students frequently attempted to apply a single operation to the numbers given.

Even when students could identify the appropriate operation to use to solve a problem, they frequently had difficulty relating the results of their calculation to the given problem in nonroutine situations. For example, the following baseball problem was administered to 13-year-olds:

> A man has 1310 baseballs to pack in boxes which hold 24 baseballs each. How many baseballs will be left over after he has filled as many boxes as he can?

Twenty-nine percent recognized that the remainder (14) of the division calculation was the correct response, but 22 percent gave the quotient (54) as their answer. This error occurred because the problem required students to do more than routinely identify an appropriate operation and perform the calculation. Apparently problem-solving involves only these two steps for too many students.

When faced with problems that contained extraneous data (see Table 5), students often attempted to incorporate all the numbers given in the problem into finding their solution. Other results showed that students did not draw pictures to help themselves understand problems (see Table 6), nor were they able to apply their knowledge of related problems to solve a given problem.

The assessment results indicate that the primary area of concern should not be with simple one-step verbal problems, but with nonroutine problems that require more than a simple application of a single arithmetic operation. Part of the cause of student difficulty with nonroutine problems may be found in our overemphasis on one-step problems that can be solved by simply adding, subtracting, multiplying, or dividing. Instruction that reinforces this simplistic approach to problem solving may contribute to student difficulty in solving unfamiliar problems. Although it may be argued that children must learn to solve simple one-step problems before they can have any hope of solving more complex problems, an overemphasis on one-step problems may teach children only how to routinely solve such problems. It may also teach them that they do not have to think about problems or analyze them in any detail.

Students need to learn how to analyze problem situations through instruction that encourages them to think about the problems and helps them develop good problem-solving strategies. There is no magic formula for making students into good problem solvers. It is clear that they need ample opportunity to engage in problem-solving activity. It is therefore important that problem solving not be regarded as secondary to learning certain basic computational skills so that students will have such opportunity.

5. Students mastered computational skills after the time of primary emphasis in the curriculum.

It is important to recognize that most computational skills are learned over an extended period of time. Assessment results suggest that most skills are mastered after their period of primary emphasis in the curriculum. For example, even though a goal of most mathematics

TABLE 5
A Problem with Extraneous Data

One rabbit eats 2 pounds of food each week. There are 52 weeks in a year. How much will 5 rabbits eat in one week?

		Percent Responding	
		Age 9	Age 13
○	2 pounds	2	2
●	10 pounds	47	56
○	52 pounds	16	5
○	104 pounds	16	11
○	520 pounds	12	23
○	I don't know	6	3

TABLE 6
Distance Problems with and without a Picture

10 ft.

6 ft.

What is the *distance all the way around* this rectangle?

		Percent Responding	
		Age 9	Age 13
○	16 feet	39	12
○	30 feet	4	1
●	32 feet	40	60
○	36 feet	4	4
○	60 feet	4	13

Mr. Jones put a wire fence all the way around his rectangular garden. The garden is 10 feet long and 6 feet wide. How many feet of fencing did he use?

		Percent Responding	
		Age 9	Age 13
○	16 feet	59	38
○	30 feet	6	3
●	32 feet	9	31
○	36 feet	5	5
○	60 feet	15	21

programs is that students learn subtraction facts by age 9, there was significant improvement in performance on subtraction fact exercises from age 9 to age 13 (from 79 percent to 93 percent), and there was also some improvement between ages 13 and 17. Similarly, students' ability to handle fraction computation increased from age 13 to age 17 (see Table 2).

Although problem solving and other content areas clearly require an increased emphasis in the curriculum, we do not deny the importance of computational skills. A reasonable level of computational skills is required for problem solving. We are suggesting, however, that problem solving not be deferred until computational skills are mastered. Problem solving and the learning of more advanced skills reinforce the learning of computational skills and provide meaning for their application.

These results also have profound implications for minimum competency programs. They suggest that rigid minimum competency programs that hold children back until they have demonstrated mastery of a given set of skills may, in fact, be depriving them of the very experiences that would lead to mastery of the particular skills.

Educators cannot be complacent, however, and assume that skills will naturally develop as students mature. Specific provisions must be made to practice and reinforce the development of critical skills. The skills that continue to develop—addition, subtraction, and multiplication of whole numbers—are skills used in a variety of contexts, so that students continue to have experiences with them in the curriculum.

Although some skills will continue to develop through use in other contexts, this is not always the case. The current high school curriculum does not take into account the fact that some students do not have many well-developed basic skills by the time they begin instruction in algebra and geometry. For example, very few 13- or 17-year-olds have mastered percentage concepts or skills; outside of general mathematics, however, there is very little opportunity for high school students to extend or maintain their knowledge of percentages.

Not only is it necessary to provide these opportunities in high school, but it is also necessary to ensure that students continue to take mathematics throughout their high school program. The assessment background data indicate that currently this is not happening. The majority of 17-year-olds take only two years of high-school-level mathematics. This situation becomes more pronounced in the case of minority students.

6. Students perceived themselves as competent, motivated, and enjoying mathematics and rated it as an important and useful subject.

The attitude exercises asked students to rate their perceptions of mathematics in a variety of ways. For example, students were asked to indicate how much they liked or disliked several subjects. In all age groups, physical education was the best-liked subject. For the 9-year-olds, mathematics was the second favorite subject (of 65 percent). While the 13- and 17-year-olds liked other subjects better than mathematics, 69 percent and 59 percent of these two groups, respectively, liked mathematics.

The self-concept statements showed that a majority of all age groups felt that they were fairly good mathematics students. Fifty-five percent of the 9-year-olds felt that they were good at working with numbers, and an additional 40 percent thought that they were sometimes good at the task. Sixty-five percent and 54 percent of the 13- and 17-year-olds, respectively, felt that they were good at mathematics, and about the same percentage responded that they enjoyed mathematics.

Over 75 percent of the 13- and 17-year-olds and 66 percent of the 9-year-olds felt that mathematics was useful in helping solve everyday problems. Further, about 80 percent of the older respondents thought that most mathematics had some practical use. Most students felt that some knowledge of mathematics was useful in getting a job. Around 90 percent of the older respondents rated mathematics as an important subject.

Students perceive mathematics as important, but as stated in the previous conclusion, the number of students continuing their mathematics declines greatly throughout the four years of high school. Although there are many reasons for this situation, two that deserve consideration come from students' perceptions of mathematics. First, in comparing mathematics to other subjects, more 13-year-olds rated mathematics easy than any other subject. By age 17, students rated mathematics hard more often than any other subject. It is necessary to examine why students' perception of the level of difficulty changes. Second, it is necessary to consider the role in which students see themselves in mathematics classrooms. The NAEP results showed that students perceived their role to be primarily passive—they are to sit, listen, and watch the teacher do mathematics, and then they are to work individually on problems from texts or worksheets. They feel they have little opportunity to interact with classmates about the mathematics they study or to explore mathematics. Not only may this role be discouraging to students, but it also may interfere with their learning. If active student involvement in the learning process is a desired goal of mathematics instruction, then changes in approaches to learning mathematics are necessary to foster and encourage that involvement.

CONCLUSION

Although many of the foregoing six conclusions have rather bleak overtones, the sixth conclusion should give us hope. Students do view mathematics positively; they see the need for it, and, contrary to some opinions, they enjoy it. During the eighties, we need to begin with the premise that our students can do better and to hold high expectations for them, as we carefully examine the ways in which we can assist them in reaching these goals.

REFERENCES

1. Carpenter, Thomas P.; Corbitt, Mary K.; Kepner, Henry S., Jr.; Lindquist, Mary M.; and Reys, Robert E. "Results and Implications of the Second NAEP Mathematics Assessment: Elementary School." *Arithmetic Teacher* 27 (April 1980): 10–12, 44–47.

2. ———; ———; ———; ———; and ———. "Results of the Second NAEP Mathematics Assessment: Secondary School." *Mathematics Teacher* 73 (May 1980): 329–39.

3. Corbitt, Mary Kay, ed. *Results and Implications of the Second NAEP Mathematics Assessment.* Reston, Va.: National Council of Teachers of Mathematics, forthcoming.

4. National Assessment of Educational Progress. *Mathematics Objectives: Second Assessment.* Denver, Colo.: NAEP, 1978.

5. ———. *Changes in Mathematical Achievement, 1973–78.* Report no. 09–MA–01. Washington, D.C.: U.S. Government Printing Office, 1979.

6. ———. *Mathematical Applications.* Report no. 09–MA–03. Washington, D.C.: U.S. Government Printing Office, 1979.

7. ———. *Mathematical Skills and Knowledge.* Report no. 09–MA–02. Washington, D.C.: U.S. Government Printing Office, 1979.

8. ———. *Mathematical Understanding.* Report no. 09–MA–04. Washington, D.C.: U.S. Government Printing Office, 1979.

THREE

Computation: Yesterday, Today, and Tomorrow
Marilyn N. Suydam

- Why should computation be taught?
- What computation should be taught?
- When should computation be taught?
- How should computation be taught?

Most teachers and parents are concerned with these questions; in fact, they have been concerned in the past and probably will continue to be concerned for some time. This article will consider some answers to these questions within the framework from which they arise—the way in which computation was viewed yesterday, is viewed today, and may be viewed tomorrow.

YESTERDAY: THE GOOD OLD DAYS

Until almost 1900, to be "great in figures" was to be learned. The mark of an educated person was the ability to compute. Most children stayed in school for only four or six or eight years (the time increasing as the century went on), learning little more mathematics than computation with addition, subtraction, multiplication, and division of whole numbers, and maybe a bit about fractions. They needed no more for their futures in farming, carpentry, surveying, shopkeeping—

or homemaking. Thus computation became synonomous with mathematics.

Throughout the twentieth century, too, even though the number of years of schooling increased and the number of career options expanded, the curriculum remained focused on computation. Computation was taught

> chiefly for its usefulness in daily life, but also because of the training that it gives the mind. (15, p. 84)

In the early 1900's, drill was the primary instructional mode: first, the teacher demonstrated a computational procedure to children, then followed it with repeated drill until the children committed the procedure to memory. This technique "trained the mind." Some drill was in the form of word problems because

> Computation is not an end in itself, but a means to an end. . . . Ability to compute is of no value unless we know what process to apply to a problem. (17, p. 13)

When the difficulties of learning purely by drill had (once again) overwhelmed teachers as well as parents (not to mention pupils), in order to promote the benefits of meaningful instruction (once again), the ultimate purpose was repeated:

> The four fundamental operations are not ends in themselves but only a means to an end; the end is problem solving. (6, p. 346)

Thus, computation and problem solving have been linked for many years.

Within the past decade, advocates of a curriculum focused on basic skills or minimum competencies have been instrumental in moving the elementary school mathematics program toward an emphasis on drill in computation. In promoting a return to the good old days, some may have overlooked not only the purpose of teaching computation, but much of reality as well.

TODAY: FACING REALITY

It has been said that the "modern mathematics" movement of the 1960's had an adverse effect on the learning of computational skills. Yet scores from many state and national assessment tests indicate that children are learning to compute—at least with whole numbers, according to reports from the National Assessment of Educational Progress. On both the first mathematics assessment in 1972–73 and the second assessment in 1977–78, performance on computation with whole numbers was generally high (2, 9). Fewer students, however, were successful with fractions, decimals, and percentages. (Also see Conclusions 1 and 3 in the preceding article, "An Interpretation of the Results of the Second NAEP Assessment.")

In a comparison of item data from a number of assessments, Bright's analysis indicated that—

1. Computational performance improves across grade levels. Computational skills are not acquired at the time of initial instruction; instead, instruction over several years is needed to reach stability, the point at which 80 percent to 90 percent attain mastery.
2. Levels of performance decrease as the items become more complex.
3. Performance tends to stabilize during the junior high school years.
4. Performance for whole-number computation stabilizes earlier and at a higher level than for fractional number computation. (1)

Bright concluded that

> It is important to note that the data presented refute the notion that students generally do not acquire basic computation skills. In fact, some skills (e.g., addition and subtraction without regrouping) are almost universally acquired, whereas others (e.g., division of decimal fractions) are not. Any meaningful discussion of the performance of students in basic computational skills must be a discussion of specific skills rather than skills in general. (1, p. 160)

Far greater difficulty was evident on word problems than on computation. (See also conclusion 4 in the preceding article.) Between the first and second NAEP assessments, problem-solving performance generally declined at all three age levels tested. In 1978, 28 percent of the 9-year-olds solved a simple word problem requiring multiplication and 46 percent solved a simple division problem, compared with 46 percent and 59 percent, respectively, in 1973. The conclusion of the NAEP summary:

> People seem to feel that facility with lower-level processes should automatically transfer into an ability to solve problems, and this is not necessarily the case. (10, p. 7)

The summary noted two points in particular:

1. Students did not seem able to think through problems.
2. Often, more students could do computations correctly than could solve word problems using the same numbers.

Several recent studies (7, 16, 20, 23) have provided evidence that over the past 25 years elementary school mathematics programs have continued to concentrate on teaching children to compute. Despite the variety of content displayed in some textbooks and curriculum guides during this period, other topics

> are most often skipped in favor of more time to develop computational skills that are comfortable to and valued by elementary teachers. (7, p. 11)

With the current emphasis on "the basics" and minimal competencies, there has been a partial return to the drill procedures common prior to 1940. Certainly recently published textbooks contain far more drill-and-practice pages than they did for some years. Certainly teachers are concerned by the pressures of accountability for student mastery of computational skills. Unfortunately, many teachers and parents do not realize that it was lack of success with rote drill programs that led directly to the proposed use of more meaningful approaches. In spite of massive effort devoted to drill, 100 percent mastery proved impossible to attain, even for drill's strongest proponents. Research indicates that meaningful instruction has a far better payoff in terms of retention and transfer (22). Memorization and drill, however, offer a payoff in immediate learning that seems enticing; moreover, drill is simpler for a teacher to administer than is meaningful instruction (for instance, using manipulative materials to develop meaning is particularly difficult for some teachers to manage).

So enticing is the promise of drill that only 14 percent of the teachers queried in a recent survey thought that mastery of basic skills should come after the development of concepts (3). Concern that a long-term problem in mathematics instruction is being worsened has led to such warnings as the following:

> Conceptual thought in mathematics must build on a base of factual knowledge and skills. But traditional school instruction far overemphasized the facts and skills and far too frequently tried to teach them by methods stressing rote memory and drill. These methods contribute nothing to a confused child's understanding, retention, or ability to apply specific mathematical knowledge. Furthermore, such instruction has a stultifying effect on student interest in mathematics, in school, and in learning itself. (7, p. 24)

Thus the needed balance between meaningful instruction and drill-and-practice procedures has not yet been realized.

Facing reality also means facing the widespread application of calculators and computers. Now in use in a vast array of occupations, these tools will be used by today's students throughout their lives. In one study, supervisors in large retail firms rated competency in their use the most important skill they wanted in hiring employees (5).

Eighty-two percent of those queried by the Priorities in School Mathematics Project (PRISM) indicated, however, that calculators should not be used until after students have learned both the meaning of whole number operations and paper-and-pencil procedures for them (13). Despite evidence that children can learn computational skills by using calculators (19), many parents and teachers do not yet accept their use.

In response to another item on the PRISM survey, over 90 percent supported the idea that paper-and-pencil computational skills should be acquired before graduation from high school. Concern for other mathematical ideas was rarely so strong. Support was very strong for one other topic, however: problem solving. It was consistently ranked high in priority for increased emphasis in the 1980's (12, p. 29). It seems apparent, then, that the beliefs about computation—carried over from "the good old days"—and about problem solving—a thread still running through today's reality—must be reconciled and merged with the calculational technology that will continue tomorrow.

TOMORROW: THE POSSIBLE DREAM

Two recommendations in the NCTM's *Agenda for Action* intertwine the roles of computational skills and problem solving and the use of technology:

1. The concept of basic skills must encompass more than computational facility.
2. Mathematics programs must take full advantage of the power of calculators and computers at all grade levels.

The rationale for such a stance notes that

> It is dangerous to assume that skills from one era will suffice for another. Skills are tools. Their importance rests in the needs of the times. Skills once considered essential become obsolete, and this is likely to increase in pace and scope as advances in technology revolutionize our individual, social, and economic lives. Necessary

new skills arise ... Time and space for including these new skills in the curriculum must be purchased by eliminating the obsolete. (12, p. 6)

The *Agenda* clearly recognizes that attaining some level of proficiency in computation without the use of calculators is necessary, but

> Common sense should dictate a reasonable balance among mental facility with simple basic computations, paper-and-pencil algorithms for simple problems done easily and rapidly, and the use of a calculator for more complex problems or those where problem analysis is the goal and cumbersome calculating is a limiting distraction. (12, p. 6)

Furthermore,

> ... even if improvement in rote computation takes place, a citizen who cannot analyze real-life situations to the point of recognizing what computations must be made to solve real-life problems has not entered the mainstream of functional citizenship. (12, p. 6)

Periodically, it is necessary to make a careful reexamination of the content of the mathematics curriculum and the way in which that content is being taught. It is also necessary to consider the four questions raised at the beginning of this article in an attempt to shape the curriculum for tomorrow.

Why should computation be taught?

Computation is a tool for solving problems in real-life situations. Thus, many persons feel that "computational skills are absolutely crucial." On the PRISM survey, 90 percent of the teachers, supervisors, mathematics educators, principals, school board members, and parents queried gave this reason for placing computation high on the list of priorities for curriculum development in the 1980's. But fewer than 2 percent felt that its importance is increasing.

As noted earlier, tradition plays a large part in shaping the curriculum. "What I learned in school is what my children should learn" is the basis for many arguments about why computational skills should continue to be taught. One factor related to this concern is that computational skill is viewed as a requirement for further mathematical study. Essentially, computation is a hurdle which students must overcome or they will be excluded from a wide range of options and occupations which require more advanced mathematics.

Those who would decrease the centrality of computation in the curriculum have responded to the foregoing reasons. They point out that computation is not used as frequently as are estimation, measurement, and other skills; they argue that, in a world in which machines can compute more accurately than people can, people should be learning to do what machines cannot do (at least yet)—think. They point out that the technological revolution is having an impact at least as great as that of the industrial revolution in the last century. Children must learn how to direct machines to solve mathematical problems, rather than spend extensive effort learning how to perform the calculations that machines can do so much more quickly and accurately than humans can.

Many persons feel that computational skills cannot be properly learned without the use of paper-and-pencil procedures. Few would eliminate such procedures, although some would advocate giving them less emphasis and eliminating those no longer needed.

What computation should be taught?

Several years ago, the National Council of Supervisors of Mathematics, concerned like many others because "computation" and "basic skills" were being equated, developed a list of ten basic mathematical skills (11). This list, endorsed by the NCTM and other educational groups, specifies the following as basic: problem solving; applying mathematics to everyday situations; alertness to the reasonableness of results; estimation and approximation; appropriate computational skills; geometry; measurement; reading, interpreting, and constructing tables, charts, and graphs; using mathematics to predict; and computer literacy. This paper places computational skills in a broader perspective:

> Students should gain facility with addition, subtraction, multiplication, and division with whole numbers and decimals. Today it must be recognized that long, complicated computations will usually be done with a calculator. Knowledge of single-digit number facts is essential and mental arithmetic is a valuable skill. Moreover, there are everyday situations which demand recognition of, and simple computation with, common fractions. (11, p. 2)

Further explication of what constitutes a desirable level of computational skill for students to attain today in order to be prepared for tomorrow has come from several sources (4, 21, 12):

1. Understanding the processes of addition, subtraction, multiplication, and division with whole numbers, fractions, decimals, and integers.
2. Memorizing the 390 addition, subtraction, multiplication, and division basic facts to the point of immediate, unaided recall.

3. Doing standard computational algorithms for addition, subtraction, multiplication, and division with whole numbers with understanding and at a moderate rate of speed, with fluency at some relatively simple computations with two to three digits (that is, up to the point where it is faster to use the head than to rely on calculators). A comparable criterion should apply to computations with fractions, decimals, and integers.
4. Developing skills in estimating, rounding, mental computation, and judging an answer's reasonableness.
5. Selecting and using computational skills in solving problems.

As Hamrick and McKillip point out, the goal should not be to turn "the student into a calculator, albeit a slow and inaccurate one" (4, p. 2). What is needed is a student who can compute without a calculator when it is more convenient to do so and, most important of all, a student who can apply computational skills in the ultimate test of solving problems. In short, students should have the computational skills they really need for tomorrow—along with other mathematical skills of vital importance.

When should computation be taught?

Answers to this question depend, of course, on what is to be taught. Not only should content be considered, but also the developmental needs of the child. Two worthwhile resources for exploring "when" are Payne (14) and Suydam (18).

How should computation be taught?

This question has been approached throughout this article. The importance of meaningful instruction, the need to place drill into its proper perspective, and the role of calculators and computers all require continued consideration.

CONCLUSION: TENETS FOR TODAY AND TOMORROW

To assist teachers in thinking about teaching computation today—and tomorrow—the following tenets are offered for consideration. It is also recommended that teachers read the entire article from which they are excerpted (21). The article discusses each tenet in detail and gives examples that are directly applicable to teaching.

1. Computational skill is one of the important, primary goals of a school mathematics program.

2. All children need proficiency in recalling basic number facts, in using standard algorithms with reasonable speed and accuracy, and in estimating results and performing mental calculations, as well as an understanding of computational procedures.

3. Computation should be recognized as just one element of a comprehensive mathematics program.

4. The study of computation should promote broad, long-range goals of learning.

5. Computation needs to be continually related to the concepts of the operations, and both concepts and skills should be developed in the context of real-world applications.

6. Instruction in computational skills needs to be meaningful to the learner.

7. Drill-and-practice plays an important role in the mastery of computational skills, but strong reliance on drill-and-practice alone is not an effective approach to learning.

8. The nature of learning computational processes and skills requires purposeful, systematic, and sensitive instruction.

9. Computational skills need to be analyzed carefully in terms of effective sequencing of the work and difficulties posed by different types of examples.

10. Certain practices in teaching computation need thoughtful reexamination.

REFERENCES

1. Bright, George W. "Assessing the Development of Computation Skills." In *Developing Computational Skills*, edited by M. N. Suydam. 1978 Yearbook. Reston, Va.: National Council of Teachers of Mathematics, 1978.

2. Carpenter, Thomas; Coburn, Terrence G.; Reys, Robert E.; and Wilson, James W. *Results from the First Mathematics Assessment of Educational Progress*. Reston, Va.: National Council of Teachers of Mathematics, 1978.

3. Denmark, Tom, and Kepner, Henry S., Jr. "Basic Skills in Mathematics: A Survey." *Journal for Research in Mathematics Education* 11 (March 1980): 104–123.

4. Hamrick, Katherine B., and McKillip, William D. "How Computational Skills Contribute to the Meaningful Learning of Arithmetic." In *Developing Computational Skills*, edited by M. N. Suydam. 1978 Yearbook. Reston, Va.: National Council of Teachers of Mathematics, 1978.

5. Mcanelly, James R. "A Study of the Mathematical Competencies Considered Important by Supervisors in Large Retail Firms in the Metropolitan Area of Chicago." Doctoral dissertation, Northern Illinois University, 1978. *Dissertation Abstracts International* 39A: 3314–15; December 1978.

6. Morton, Robert Lee. *Teaching Arithmetic in the Elementary School.* New York: Silver Burdett, 1937.

7. National Advisory Committee on Mathematical Education, *Overview and Analysis of School Mathematics, Grades K–12.* Washington, D.C.: NACOME, Conference Board of the Mathematical Sciences, 1975.

8. National Assessment of Educational Progress. *Changes in Mathematical Achievement, 1973–78.* Report no. 09–MA–01. Denver: NAEP, August 1979.

9. ———. *Mathematical Knowledge and Skills.* Report No. 09–MA–02. Denver: NAEP, August 1979.

10. ———. *Mathematical Achievement* (Summary). Denver: NAEP, 1979.

11. National Council of Supervisors of Mathematics. "NCSM Position Paper on Basic Mathematical Skills." January 1977. ED 139 654. (Also in *Arithmetic Teacher* [October 1977] and *Mathematics Teacher* [February 1978].)

12. National Council of Teachers of Mathematics. *An Agenda for Action: Recommendations for School Mathematics of the 1980s.* Reston, Va.: The Council, 1980.

13. ———. *Priorities in School Mathematics: Executive Summary of the PRISM Project.* Reston, Va.: The Council, 1981.

14. Payne, Joseph N., ed. *Mathematics Learning in Early Childhood.* Thirty-seventh Yearbook. Reston, Va.: National Council of Teachers of Mathematics, 1975.

15. Smith, David Eugene. *The Teaching of Arithmetic.* Boston: Ginn, 1909.

16. Stake, Robert E., and Easley, Jack. *Case Studies in Science Education.* Final Report, National Science Foundation Contract No. C 7621134. Urbana, Ill.: University of Illinois, 1978. ED 156 498–513.

17. Stone, John C. *The Teaching of Arithmetic.* New York: Benjamin H. Sanborn and Co., 1918.

18. Suydam, Marilyn N., ed. *Developing Computational Skills.* 1978 Yearbook. Reston, Va.: National Council of Teachers of Mathematics, 1978.

19. Suydam, Marilyn N. *Using Calculators in Pre-College Education: Third Annual State-of-the-Art Review.* Columbus, Ohio: Calculator Information Center, August 1980.

20. ———, and Osborne, Alan R. *The Status of Pre-College Science, Mathematics, and Social Science Education: 1955–1975. Volume II: Mathematics Education.* Final Report, National Science Foundation Grant No. NSF–C76–20627. Columbus, Ohio: ERIC Clearinghouse for Science, Mathematics, and Environmental Education, 1977. ED 153 878.

21. Trafton, Paul R., and Suydam, Marilyn N. "Computational Skills: A Point of View." *Arithmetic Teacher* 22 (November 1975): 528–37.

22. Weaver, J. Fred, and Suydam, Marilyn N. *Meaningful Instruction in Mathematics Education.* Columbus, Ohio: ERIC Clearinghouse for Science, Mathematics, and Environmental Education, March 1972. ED 068 329.

23. Weiss, Iris. *Report of the 1977 National Survey of Science, Mathematics, and Social Studies Education.* Final Report, National Science Foundation Grant No. C76–19848. Research Triangle Park, N. C.: Research Triangle Institute, 1978. ED 152 565.

FOUR

Measurement Is Basic

Gwen Shufelt

There is little question about the significance of measurement as basic to the realization of a career in such areas as medicine, engineering, computing, or applied sciences. Measurement is such an integral part of everyone's daily experience, however, that it is often overlooked as a basic skill in mathematics education. It would be difficult to imagine a day without some measurement reference. At the most trivial level, questions of this nature are posed: Have I gained or lost weight? Is it cold enough to wear a sweater? Do I have time to finish this task before my favorite television program?

Even these simple applications involve reading appropriate measurement units on the bathroom scale and the thermometer and computing lapsed time on the clock. All assume familiarity with units appropriate to the property being measured—weight, temperature, time.

Estimations of length, area, and space, with or without reference to standard units, are another basic part of daily experience. Can I get my car into that parking space? Is that rug the right size for my room? If I buy this new cabinet will it fit the space I have?

At a slightly higher level of application, such daily activities as cooking, sewing, home improvement projects, or a garden plan involve more precise uses of measurement. Thus, basic to minimal functioning in everyday activities is competency with both the instruments and units related to obtaining commonly used measurements.

DEVELOPMENT OF MEASUREMENT CONCEPTS

In the primary grades, the concept of measurement as a comparison should at first be developed without reference to particular units. Direct comparison of two objects that have the property under consideration should be made. Is one child taller than another? Compare by having children stand side by side (or back to back). Is the orange heavier than the apple? Compare by using hands to "feel" weight or by using a balance to "see" which one is heavier.

Later the measurement of objects that cannot be brought together for comparison can be achieved by the use of an intermediary, movable object that has the property to be measured. For example, to compare the width of my desk at school with the width of my desk at home, I can use my new pencil. One desk is about four pencils wide. The other is about five pencils wide. Not only do I know which is wider, I develop the measurement concept of repeating a unit, my pencil, to assign a number to a continuous property such as the width of a desk.

Teachers should be aware of the conceptual difficulties inherent in the use of number in a measurement context as different from cardinal (or even ordinal) uses relative to discrete objects. Perhaps because of early socialization and instruction a child's concept of number seems to be biased toward number associated with discrete objects or even used in a nominal sense rather than number assigned to a continuous property. Children often learn to count and to identify number associated with the house or apartment "where I live" even before coming to school. Perhaps there is a readiness in terms of Piagetian conservation that must be realized, but there may also be the factor of lack of experiences with number in a measurement sense.

From the beginning, the development of a child's concept of measurement should emphasize measurement as an approximation. Words such as "about four pencils" long or "to the nearest unit" should be a part of the basic vocabulary of measurement. Furthermore, emphasis should be given to the convenience of subdividing units to obtain smaller units for measuring smaller objects or creating larger units (multiples of a basic unit) to measure larger objects. Moreover, the ability to select an appropriate unit, both in terms of size and characteristics, must be developed as a component of basic measurement skills.

In the middle grades, the need for standard units may be motivated out of the limitations of nonstandard units. My hand and your hand may not be the same size. Which hand should be used to measure the table? If we are to communicate our measurements, units that are both standard and universal must be learned.

Thus, early instruction in measurement should include activities using both nonstandard and standard units. The purpose of the unit should be emphasized. It must have the property being measured and be appropriate in size. The process of measurement then is developed as a repetition of the unit until a "best fit" is achieved and the approximate number of units for a particular measurement is obtained. If a child really comprehends this process, then the learning of a particular standard system of measurement merely involves familiarization with the units of the system chosen.

Traditionally, at the secondary level, the teaching of measurement has been delegated to the science department. It has been taught predominantly in an application setting with scant attention to the mathematical subtleties involved in the concepts of precision, greatest possible error, and absolute error.

These topics are entirely appropriate for a mathematical treatment as applications of rounding and intervals. If measurement topics could be developed cooperatively between mathematics and science departments, students would benefit from studying the mathematical theory of measurement independent of any system of standardized units in the mathematics class. Then the science laboratory would provide the applications with a specific system of units—the metric system.

Measurement as an approximation provides an appropriate setting for the introduction or clarification of the concept of a real number as a nested interval on the line. Linear measurement associates a real number with the length of a segment relative to a unit on a number line. Greatest possible error provides an interval within which lies the number that is the length of the segment. Furthermore, the concept of precision can be made explicit with reference to the size of the unit used. These topics are appropriate for consideration in the secondary mathematics class.

THE METRIC SYSTEM

Unfortunately, today children in this country are caught in the changeover from the traditional system, based on English units, to the international metric system. Too many of the adult population, outside the scientific and industrial sectors, seem to be clinging tenaciously to the cumbersome yards, feet, inches, pounds of the English system in spite of federal legislation intended to move the country toward universal use of the metric system.

Both systems appear on many standardized tests. Both are included in the majority of elementary text series. So it seems that children will continue to be instructed in both for some time to come. This situation should not pose a problem to the elementary teacher if there is strict adherence to a policy of eliminating any formal conversion from one system to another. Rather, the emphasis should be on familiarization with the units of each system independently. This familiarization should come to students through the actual use of measuring devices graduated in the units of the system. Measurement is one basic skill that can best be learned through activities and experience using the appropriate units and tools. It cannot be learned solely as a textbook subject. This applies to the learning of both traditional and metric units.

In the early grades, as soon as the need for standard units has been demonstrated, students should be introduced to the basic linear, weight, and capacity units for the metric system. At this stage the meter, decimeter, and centimeter should be used to measure lengths until a child is familiar enough with the units to use them to estimate lengths. Gram and kilogram weights should be used to balance familiar objects on a balance scale to obtain the same familiarity. At this stage, relationships between units in the system should only be developed out of counting how many centimeters in a meter, not by analysis of prefixes (e.g., centi- means 0.01 meter). That activity should be delayed until students have developed base ten concepts associated with multiplication and division by powers of ten.

If English units are to be introduced (and children will probably continue to encounter their use at home for some time), they should be introduced in much the same way as suggested for metric units. Children should use the units in actual measuring activities. They should learn the relationships between units from observing and counting on a ruler or on a scale.

Through upper elementary grades and into middle school grades, students should extend their knowledge of metric (and English) units to all units that they will need in everyday activities. These should include kilometer and millimeter in addition to the linear units previously taught. Liter and milliliter as capacity units should also be taught.

By middle school age, most students should be ready for a more formal treatment of prefixes in the metric system as related to the basic unit in each measurement area. They should master at least the following common prefixes and basic units:

Prefixes: milli-, centi-, deci-, and kilo-.
Basic Units: the meter as the unit of length; the gram as the unit of mass; the liter as the unit of capacity; the Celsius degree as the unit of temperature; and of course time units.

With this background in elementary and middle school, high school students who pursue academic programs in mathematics and science should be much better prepared to be successful in the more abstract formulations of measurement they will encounter. At this level complete formalization should be achieved, and the entire instructional emphasis should be on mastery of all metric system units. The metric system as the language of science should be the only system treated in high school courses. The old practice of spending time in science courses converting from metric to English and English to metric units should be abandoned completely. Modern high school students should live in a metric world.

Even those students who will not continue with additional formal study after high school should be provided with enough continuing instruction in metric measurement to enable them to be competitive for jobs. Many technical and business occupations will soon be totally metric. This includes such changes as the use of metric tools by mechanics, packaging and bottling in metric units, and many other changes.

GENERALIZATIONS AND FORMULAS

Measurement of area and volume should be introduced first as "covering" and "filling" with appropriate units, square units for area and cubic units for volume. Students should construct these units from familiar linear units and should obtain area and volume measurements by "filling" and counting. Such experiments should precede discovery of the generalizations that lead to formulas.

For example, given centimeter-squared paper a child should first cover a rectangular region with squares and then count the number of squares to find the area. Then the fact that the length gives the number of squares in a row and the width gives the number of squares in a column leads the student to a shortcut (formula) for finding the area: $A = l \times w$ (see Figure 1).

Similarly, by first counting squares and then relating the region enclosed by a parallelogram to a rectangular region with equal area, students can develop the formula for the area enclosed by a parallelogram (see Figure 2). Also, two congruent triangular regions may be joined to form a parallelogram and the area of the triangle shown to be one-half the region enclosed by the parallelogram (Figure 3).

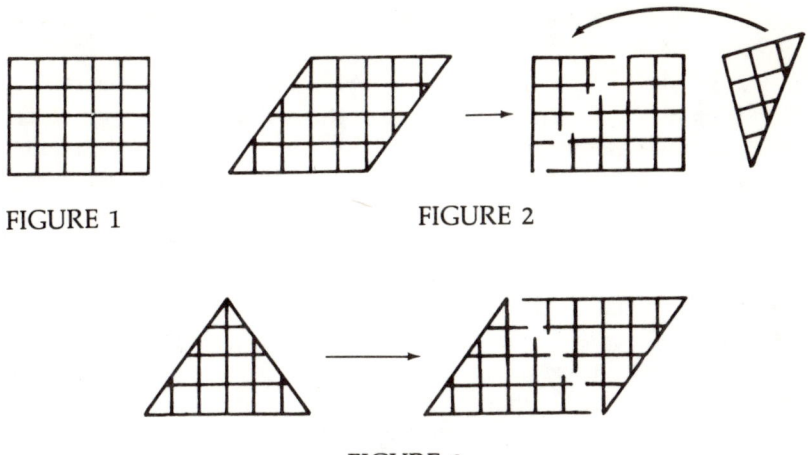

FIGURE 1 FIGURE 2

FIGURE 3

The development of formulas for measurement of area in this manner helps give meaning to them. It also avoids the confusion of area and perimeter that often occurs when the formulas are introduced without experiments to make them understandable.

Even the area of a circular region can first be explored by covering it with squares, counting and estimating the area. Approximations for pi can be obtained by measuring the radius, squaring and dividing into the estimated area. Careful experimenting with several circles of different diameters can lead to the discovery of pi as the constant ratio of circumference to diameter (a number a little greater than 3). This comparison of circumferences with diameters of many circles leads to generalization of the formula $C = \pi d$.

Such exploratory activities should be used to introduce each of the measurement concepts including volume, capacity, mass (or weight), temperature, and time. Insofar as possible, actual use of instruments, reading scales or counting units, making comparisons should precede any attempt to generalize with formulas. The added benefit is that students will be prepared to use the tools and standard units of measurement in daily applications.

CONCLUSION

In a concern for "back to basics," measurement as a basic skill for everyone must receive the attention of all educators. As the technological revolution continues, the uses cited in this article are only an infinitesimal part of the potential role of measurement in the life of every individual.

Minimal literacy will require the ability to use and understand a variety of measurement units. Even at the ordinary consumer level, measurement comparisons are essential to wise spending. As inflation continues, such consumer wisdom becomes even more critical.

Similarly, beyond minimal competency, if this country is to continue to prosper in an international setting, education must provide the scientists and the engineers to maintain a competitive technology and industry. An essential ingredient in such education is a thorough foundation in all aspects of measurement.

Between these extremes, minimal daily existence and top-level technological contribution, lie many tiers of varying sophistication in the applications of measurement to everyday living and to jobs. Measurement is indeed a basic skill.

FIVE

Computational Estimation Is a Basic Skill

Robert E. Reys and Barbara J. Bestgen

Computational Estimation Is a Basic Skill

COMPUTATIONAL ESTIMATION

Did you try to do these problems quickly in your head? Most people do. It is not always convenient to write everything down. In most cases an exact answer is not necessary but the answer should be close enough to allow for whatever decisionmaking is required. For example, if you have only one dollar but order a hamburger and milk, this decision would lead to some embarrassment. Everyone encounters such consumer-oriented problems daily. Each problem involves computational estimation and thus reflects the theme of this article, namely that *estimation is a basic skill.*

All of the opening questions require problem-solving skills. However, each of them is different from typical school mathematics problems. For example,

1. They can be solved without paper and pencil.
2. They rely on mental computation.
3. They can be done quickly since time is usually at a premium.
4. They result in answers that are not exact but are adequate for making necessary decisions.

These features characterize computational estimation, which has been identified as a basic skill (1, 5, 7). In fact, mental computation and estimation skills are used much more frequently than paper-and-pencil algorithmic procedures in solving real-world problems involving mathematics. Whereas technological advances with calculators and microcomputers have lessened the need for traditional paper-and-pencil algorithms, they have increased the need for computational estimation skills that provide quick checks of the reasonableness of answers.

STATE OF THE ART

Despite its importance, estimation is perhaps the most neglected skill in the mathematics curriculum (3). Traditionally introduced around the fourth grade, computational estimation frequently appears as a separate topic that is poorly motivated and often ignored in later work with computation. A review of mathematics basal textbooks (10) shows very little attention given to the systematic development of computational estimation skills. In the vast majority of series the emphasis given is woefully lacking. According to a recent study of three popular mathematics textbook series, estimation appeared in less than 3 percent of the lessons (6).

This lack of attention to computational estimation in school programs was documented in a recent study of in-school secondary students and out-of-school adults who had been identified as good computational estimators (8). In answer to the question "Have you been taught how to estimate in school?" the predominant answer was that the students had been taught to round numbers, but they rarely used this skill in conjunction with either the development or practice of estimation ability. Most students voiced uncertainty about where or how they had developed their skill, frequently suggesting that they picked it up through the need for an efficient, reasonably accurate computational tool. Conversations with the adults provided similar information; they could not recall estimation being explicitly taught in school.

This lack of attention to computational estimation is reflected in the low performance of all age groups in the second National Assessment of Educational Progress (4). The results of one multiple choice exercise vividly illustrate the poor performance of students:

ESTIMATE the answer to $12/13 + 7/8$. You will not have time to solve the problem using paper and pencil.

Choices	Age 13	Age 17
◯ 1	7%	8%
◯ 2	24%	37%
◯ 19	28%	21%
◯ 21	27%	15%
◯ I don't know	14%	18%

These results show that only 24 and 37 percent of the 13- and 17-year-olds, respectively, responded correctly. Even worse is the fact that over one-half and one-third of the two age groups selected values that were completely unreasonable. Rather than estimate the sum, many students attempted to operate directly on the numbers with no concern for the reasonableness of their estimate. These performances were consistent with those reported by the National Longitudinal Study of Mathematics Ability (11).

There has been little research on the ability of students to perform computational estimation. When students are asked to estimate, their response is often to try to work the problems quickly with paper and pencil, then round their answer to reflect an "estimate" (2). This procedure is clearly not computational estimation, but the tendency to use it is a major confounding variable in assessing estimation. This is one of the reasons why estimation skills are difficult to assess and have discouraged the authors of many standardized tests from including questions related to computational estimation. In fact, the statewide assessment in Missouri, the *Basic Essential Skills Test* (BEST), explicitly states that estimation is viewed as a basic mathematical skill but that the responsibility for checking student performance rests at the local school level. The irony of a test purporting to measure basic skills but not attempting to assess computational estimation is difficult to understand. Nevertheless, it is a testimony to the psychometric problems that are introduced when assessing computational estimation (9).

A recent research study identified and described computational estimation processes used by good estimators (8). This research resulted in identifying some specific characteristics, skills, and thinking strategies that have instructional implications. The instructional suggestions that follow reflect this research base. It is hoped that these ideas will stimulate increased efforts to develop computational estimation throughout the mathematics program.

POWER AND PURPOSE OF ESTIMATION

A major obstacle to instruction on estimation can stem from inappropriate attitudes about what estimation is and its potential use. Students see estimates as "second-rate" answers, not quite so good or useful as exact answers. This attitude can cause students to ignore techniques being taught and simply try to compute faster. For this reason, an awareness of the importance of estimation can and should begin with young students. Instruction and experiences in generating real-world estimates will further enhance students' awareness of the power and usefulness of estimation.

Beginning with a discussion of various examples of estimates, teachers should emphasize the everyday, real-world importance of estimation. For example, which of the following situations expect an estimate? Which require an exact answer?

How much money will be needed for the Saturday afternoon show?
How old are you?

How many brothers and sisters do you have?
What is the time?
How many people attended Friday's football game?
What tip should I leave the waitress?

For some situations an exact answer is essential and an estimate makes little sense. In other instances, such as one's age, an exact answer is cumbersome and provides no better information than an estimate. Students quickly realize how often estimates are used in their daily experiences and can begin to generate other examples. Many terms are associated with estimates. These terms help us recognize an estimate. Encourage students to generate a list such as the following: almost, nearly, close to, approximately, around, over, about.

These initial experiences will help students understand the power and purpose of estimation. Once this attitude starts to form, instruction on a variety of strategies can begin.

FRONT-END STRATEGY

A very basic, yet powerful, estimation strategy which can be used in a variety of situations and taught to the young student is a front-end strategy. In estimating, the most important digits in a number are the leading or front-end digits. Unlike computing with paper-and-pencil algorithms where work often begins on the back-end digits, an estimate requires a quick and accurate answer that is arrived at most efficiently by focusing on the front-end digits. To help students understand this idea, hide a three-digit number behind a poster on the board. Ask students to guess a three-digit number and see how close they can come to the hidden number. Before guessing, give them the opportunity to see one digit of their choosing. As students try this activity, some will choose the hundreds digit. This gives the most useful information. Others will ask to see the ones digit. Will this information be of as much help? Why? A few such experiences will help students see that the leading digit is powerful because, together with its place value, it represents a good approximation of the original number.

To introduce the front-end strategy for addition, present a problem such as the one that follows. A portion of the problem has been torn away and some information is missing. If the missing piece cannot be found, can an exact answer be determined?

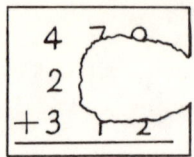

Since the leading digits are visible, students can use them to formulate an estimate. For example, $4 + 2 + 3 = 9$. Is 9 a good estimate? It looks as if each of the numbers in the problem contains 3 digits, therefore 900 rather than 9 is a reasonable estimate. Other discussion can touch upon such questions as these:

Is 900 an over- or underestimate?
How can we get a closer estimate?

Practice in using this strategy is important. Initially, provide the cover for the back-end digits as in the next example. Give students similar problems where they hide the back-end digits by covering them with a hand or card. It is also helpful to limit leading digit addends to a sum of nine or less. After a few such experiences, pose an example where this sum causes an increase in digits.

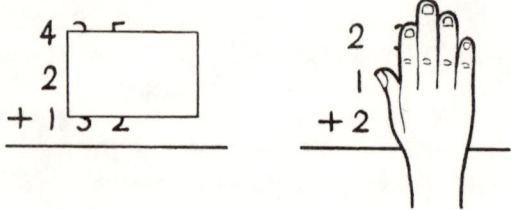

Immediate feedback on students' performance in providing estimates is important. Be lenient in accepting responses initially, but ask students to explain how they obtained their estimate. Their explanations will help clarify procedures and lead to greater understanding of their processes, and also suggest new approaches to estimating a given problem.

Although we have suggested the use of this strategy with young students, there are a variety of applications of its use for older students and adults. For example,

What is the attendance for the past five Soap Bowl games?	1978	42,946
	1979	51,895
	1980	48,987
	1981	71,432
	1982	78,823

Front-end
Solution: 4 + 5 + 4 + 7 + 7 gives 27, and about 20 thousand more is about 290,000.

In using this strategy, the focus is on two important aspects of formulating an estimate: the leading digit and the place value of that digit. Other followup activities might include:

Using larger numbers and/or more addends. Presenting problems where the place value of the numbers is different.

```
  436
   79
 +204
```

Presenting problems in a horizontal format. Can students find the important digits?

213 + 46 + 193

Developing the front-end strategy with subtraction. What different ideas are involved?

```
  4236
 -2517
```

Although the front-end strategy illustrated is easy to understand and apply, more precise strategies exist and should be developed as students mature. For example, if each addend of

```
  436
   79
 +204
```

were rounded to the nearest hundred, 400 + 100 + 200 gives a good estimate of the sum. This technique requires both rounding and mental visualization of the numbers, which adds to the complexity of the task. The added precision yielded by this technique is important, however, and should be encouraged as development continues.

USING COMPENSATION TO REFINE ESTIMATES

Following development of mental computation techniques with multiples of 10, estimation of products follows naturally. For example,

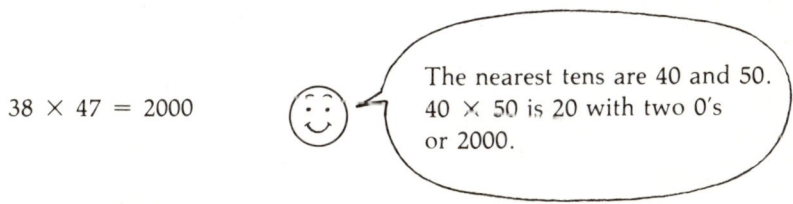

38 × 47 = 2000   The nearest tens are 40 and 50. 40 × 50 is 20 with two 0's or 2000.

38 × 47 = a little under 2000

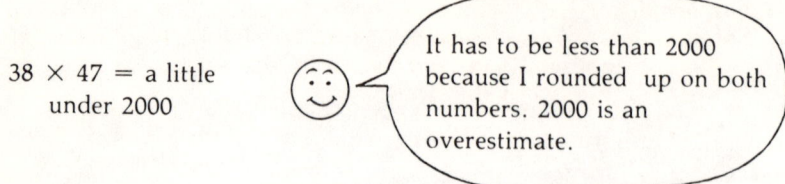

It has to be less than 2000 because I rounded up on both numbers. 2000 is an overestimate.

The student in this example uses the idea of compensation to further refine the initial estimate. What about these examples:

	Think	Reported Result
42 × 61 =	40 × 60 = 2400	"It is a little more than 2400."
39 × 78 =	40 × 80 = 3200	"It is a little less than 3200."
27 × 32 =	30 × 30 = 900	"It is about 900."

For the first two examples, the rounding procedure makes it clear that the reported result is an underestimate or an overestimate. What about the last example? It is not obvious without further exploration.

SOME FINAL THOUGHTS AND SUGGESTIONS

Specific estimation strategies abound. As students begin to practice and use estimation, they will develop strategies and techniques for different operations and types of numbers on their own. Verbalization of these specific strategies will help clarify and refine their use and will encourage others to use these ideas and to search for and discover other ways to efficiently estimate. The importance of such discussion cannot be overemphasized.

Guiding students through a meaningful and lasting development of estimation strategies and subskills cannot be done in a single unit. Instruction should be systematic, starting as early as third grade. We have described several suggestions for specific lessons. Here we highlight some additional suggestions and ideas for implementing an estimation learning sequence.

1. Instruction on estimation should include discussions of why and when to use this skill over exact computation. This discussion will help students develop a tolerance for error so that they can begin to feel com-

fortable with an answer that is not exact. For example, a review of a few articles in a daily newspaper will reveal many examples of numerical data—some of which are estimates, some exact values. Have students identify the examples they feel are estimates and discuss why.

2. Strategies such as the front-end technique described can be meaningfully developed with young students. Therefore, selected strategies should be taught early and not delayed until later grades.

3. Mental computation is an important prerequisite skill for many estimation strategies. Therefore, this skill should be systematically developed and practiced prior to and along with estimation. A starting point is work with powers and multiples of 10. What patterns appear when powers of 10 are multiplied by numbers? What patterns appear when multiples of 10 are multiplied by other multiples of 10? These ideas can be developed and refined by providing a few experiences. For example,

x	10	100	1000
3			
4			
7			
9			

x	30	50	60
20			
40			
50			
70			

What patterns emerge as these tables are completed? Can students verbalize rules for multiplying by powers of 10? How about multiples of 10? This work will help pave the way for success at using rounding and other more sophisticated strategies.

4. Instruction on how place value is affected by certain operations of arithmetic will help correct order of magnitude errors for multiplication and division. For example,

$$59 \overline{)472{,}869}$$

The estimator might reformulate the problem to $60 \overline{)480{,}000}$. Is the estimate 8? 80? 800? 8,000? or 80,000? Instruction that includes techniques for determining appropriate place value will help avoid unreasonable answers.

5. The idea of compensation—that is, refining an initial estimate—should be encouraged. Verbal compensation can be used in the beginning. This refers to stating whether an estimate is "a little over" or "a little under" the exact answer. Later, arithmetic procedures can be used to establish a numerical compensation. For example, to estimate 27 × 41, we could round to 30 × 40 or 1200. 1200 is about three 40's too high, so a more refined estimate is 1200 − 120 or 1080. Applied estimation problems may evoke a different type of compensation. For example, in the following problem an overestimate is likely to occur to ensure having enough money to attend the program. Thus, in this example, a meaningful rationale can be given for compensating upward.

> Movie tickets cost $3.25 for adults and $1.75 for children. About how much will tickets for three adults and one child cost?

6. The value of applications of estimation cannot be overemphasized. Real-world consumer-related and other settings should be used to develop and practice estimation strategies. For example, supply a grocery store sale bulletin and estimate the cost of grocery lists. Other uses of this bulletin can be made for generating estimation questions.

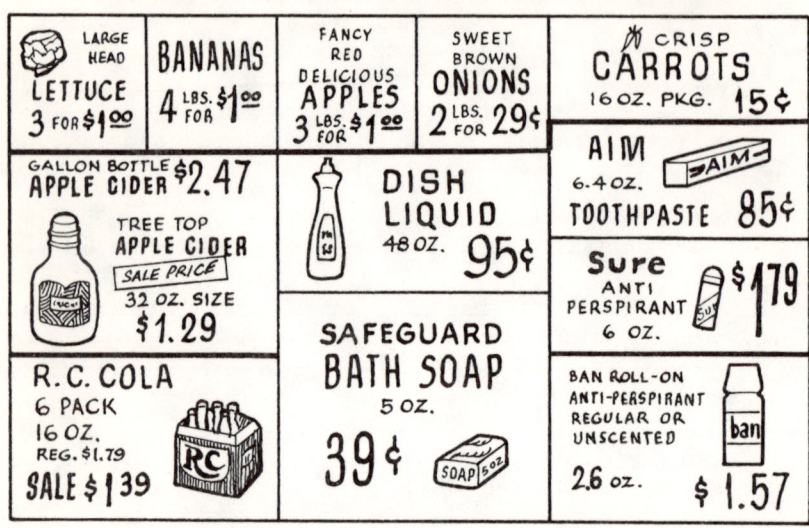

About how much will five bars of Safeguard soap cost?
Is the dish liquid priced at more or less than 2¢ an ounce?
About how much will five pounds of sweet brown onions cost?

A school activity such as a play can also generate estimation questions:

About how many people will we be able to seat in our gym?
If we sell advance tickets, how many tickets will each member of the class need to sell to fill the gym?
If we charge 50¢ for children and 75¢ for adults, what will be our income?
About how much will our expenses be?
How much profit can we expect?
If we sell cokes for 30¢ each during intermission, how much will we make? (To answer this, we'll need to know about how many people will likely buy a coke.)

7. Finally, instruction on estimation should be given on a regular basis. Regular practice and feedback should occur frequently (perhaps two 10-minute sessions weekly). A quick practice idea is to take three or four pieces of tagboard and place a number on each. For example,

| 347 | 10,427 | 6,819 | 4,709 |

Ask students to number their papers from 1 to 4 (or more). Show students any two (or three) cards at a time and ask them to estimate the sum (or difference). From these four cards you can generate six different two-addend problems. You can also easily control the time allowed on each problem.

NO DEFAULTS, PLEASE

We have presented an overview of computational estimation, which is indeed a basic skill. Although current performance levels are disappointing, they seem to reflect the attention given this topic in current mathematics programs. According to experience and research, useful estimation techniques exist but are possessed by very few. In order that more people may develop these skills, computational estimation must be integrated into current mathematics programs and must be taught systematically. Research has provided new insight and some direction. In this discussion we have provided a modest but essential first step that must be taken to get started. So little has been done with computational estimation, that, in a very real sense, it is a new game. The ball is in the reader's court. The importance of keeping it in play should not be underestimated.

REFERENCES

1. Bell, M. S. "What Does 'Everyman' Really Need from School Mathematics?" *Mathematics Teacher* 67 (March 1974): 196–202.

2. Bestgen, B. J.; Reys, R. E.; Rybolt, J. F.; and Wyatt, J. W. "Effectiveness of Systematic Instruction on Attitudes and Computational Estimation Skills of Preservice Elementary Teachers." *Journal for Research in Mathematics Education* 11 (March 1980): 124–36.

3. Carpenter, T. P.; Coburn, T. G.; Reys, R. E.; and Wilson, J. W. "Notes from National Assessment: Estimation." *Arithmetic Teacher* 23 (April 1976): 297–302.

4. ————; Corbitt, M. K.; Kepner, H.; Lindquist, M. M.; and Reys, R. E. "Results and Implications of the Second NAEP Mathematics Assessment: Elementary School." *Arithmetic Teacher* 12 (April 1980): 10–12+.

5. National Institute of Education. *NIE Conference on Basic Mathematical Skills and Learning* (Euclid, Ohio, October 4–6, 1975). 2 vol. Washington, D.C.: The Institute, 1975. ED125 905 and ED125 909.

6. Freemen, D.; Kuhs, T.; Belli, G.; Floden, B.; Khappen, L.; Porter, A.; Schmidt, B.; and Schwille, J. "The Fourth Grade Mathematics Curriculum as Inferred from Textbooks and Tests." Paper presented at 1980 Annual Meeting, American Educational Research Association, Boston, Mass.

7. National Council of Supervisors of Mathematics. "Position Paper on Basic Mathematical Skills." Minneapolis: The Council, 1977.

8. Reys, R. E.; Bestgen, B. J.; Rybolt, J. F.; and Wyatt, J. W. "Identification and Characteri-

zation of Computational Estimation Processes Used by Inschool Pupils and Out-of-School Adults." National Institute of Education Grant No. 79–0088, November, 1980.

9. ——, and ——. "Computational Estimation: What Needs to Be Done and How to Get Started." *Elementary School Journal,* in press.

10. Skvarcius, R. "The Place of Estimation in the Mathematics Curriculum of the Junior High School." In *Cape Ann Conference on Junior High School Mathematics.* Boston: Physical Science Group, 1973.

11. Wilson, J. W.; Cahen, L. S.; and Begle, E. G., eds. *NLSMA Reports* (1A, 2A, 4, 5). Palo Alto, Calif.: School Mathematics Study Group, 1968.

SIX

Finding and Using Data: A Basic Skill
Albert P. Shulte

Nearly everyone needs the ability to deal with data. In a typical day, an average person might do the following: look at a table of ingredients to prepare a new recipe; read a mileage table to figure out the distance to another city; look at the batting averages in the sports page to check the hitting of local favorites; read the results of a poll on preferences in a gubernatorial race; look at a graph to see how much a tax limitation proposal will save taxpayers; examine a chart or profile of a child's performance on a standardized test.

At the same time, the schools are presently doing little to improve students' ability to deal with data. Part of the problem is the name often attached to this skill—the word *statistics* scares many people. Also, mathematics instruction at the elementary school level tends to concentrate almost exclusively on computation. And, at the secondary school level, work with data is usually included as an optional chapter late in the textbook, and thus is treated as an expendable extra.

What can be done to provide students more opportunities to work with data? In this article, a number of activities at different grade levels will show ways to provide such opportunities, without requiring large blocks of time, much technical language, or extensive background on the part of the teacher.

In this article, data-related experiences are classified into two major groups: (1) *finding* data—sources of data that can be used with students; (2) *using* data—what can be done with data after they are collected.

FINDING DATA

Data may be collected from a variety of sources. The teacher who wishes to use data in the classroom and who keeps an eye open for interesting items will discover many good sources in his/her daily reading. The newspaper is a prime source. It uses graphs to present information quickly. It includes tables to provide background for articles. It frequently describes methods of sampling. Feature articles of the day often include the results of polls (formal and informal). The sports pages are full of statistics.

Another source is the magazines and periodicals people read for subjects of personal interest. For example, two of my favorite examples came from *Natural History* magazine. One article discussed what causes flying squirrels to store nuts—a fine controlled study presented by means of graphs. Another article compared roadrunners to pigeons in various tests to see why roadrunners were specially adapted to live in the desert (it turned out that they were not). The *TV Guide* is an excellent source—ratings are so important in television that many articles appear on subjects such as the following: Which programs appeal most to young adults? Does TV make children more violent? Why does golf, watched by relatively few people, rate so much exposure?

A number of general references can supply data of interest to students. For example, *The Book of Numbers* is a fine source. Among other references, the *Guinness Book of World Records* can be used as the basis for contests in which students can generate their own data; and the *Statistical Abstract of the United States* contains a wealth of data, some of which is interesting to students.

Students can also be encouraged to collect their own data. They can bring in items from their own reading or browsing. They can be assigned to look through newspapers for data, or to examine magazines in the school library for appropriate articles. They can perform experiments and record their results. They can carry out surveys, which requires them to deal with such questions as sampling error, biased samples, taking random samples, and making sure that the sample is representative of the group being sampled.

USING DATA

Once data are collected, what can students do with the information? They can learn to *organize* the data, to *display* the data effectively, to *summarize* the data, to *make predictions* based on what they have found, and to *make inferences*. Here are some examples to illustrate these areas.

Organizing and Displaying Data

Object Graphs. Young children can make graphs by using actual objects and attaching them to sheets of tagboard. For example, they can show their preferences in color of crayon by attaching the appropriately colored crayon to the tagboard.

Picture Graphs. After experience with object graphs, students can move to the next level—using pictures or drawings of objects and attaching these to tagboard. For example, they can use pictures of their favorite pets. Another graph appropriate for primary classes is one with a paper tooth attached for each tooth that a child has lost.

TEETH MISSING	
Sally	🦷🦷🦷
Wendell	🦷
Trena	🦷🦷
Lupe	🦷🦷🦷

Tallying. Students can be taught to tally in the usual manner, showing the results of throwing a die, spinning a spinner, or some other activity. The following illustration shows the results of three children playing a game 20 times.

STUDENT	NUMBER OF WINS
Sally	ℍℍ III
Trena	IIII
Wendell	ℍℍ III

Stem-and-Leaf Display. Upper elementary and middle-school students can be shown how to present information using this new technique. Suppose the class has 29 students, and they have found their heights in centimeters: 157, 148, 160, 142, 151, 164, 151, 146, 134, 138, 162, 142, 155, 155, 151, 167, 152, 143, 159, 154, 157, 140, 155, 142, 145, 159, 138, 141, 146. These data are arranged in the following stem-and-leaf display, with the first two digits the *stems* and the third digits the *leaves.*

```
              13 | 4, 8, ⑧——— LEAF
   STEM ——⑭  | 8, 2, 6, 2, 3, 0, 2, 5, 1, 6
              15 | 7, 1, 1, 5, 5, 1, 2, 9, 4, 7, 5, 9
              16 | 0, 4, 2, 7
```

Notice that none of the original information is lost, and the arrangement of the data makes a sort of bar graph. If the stems are arranged in numerical order, the middle score (the *median*) can be easily located as follows:

```
13 | 4, 8, 8
14 | 0, 1, 2, 2, 2, 3, 5, 6, 6, 8
15 | 1, ①, 1, 2, 4, 5, 5, 5, 7, 7, 9, 9     MIDDLE HEIGHT
16 | 0, 2, 4, 7                              (MEDIAN)
                                             (151 Cm)
```

Box-and-Whiskers Plot. This technique can be used from middle school up to give a picture of a collection of data. Using the height data above, students can find the *quartiles.* These are the middle scores in the upper half and the lower half of the scores: 157 cm and 142 cm, respectively. The box-and-whiskers plot follows.

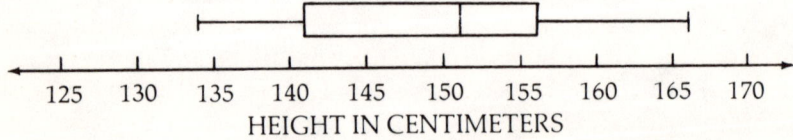

The box shows the middle half of the class. The whiskers go from the box to the shortest student, 134 cm, and to the tallest student, 167 cm.

The median is shown as a vertical line inside the box.

Summarizing Data

Averages. Two types of numbers are important in summarizing data. One of these is a number that gives some information about the middle of the distribution—an average. The three most common types of averages are as follows: the *mean* (which is what most people call the "average" —add all the numbers and divide by the number of numbers); the *median* (the middle score in the distribution); the *mode* (the most common number). In the example of heights, the median has already been given as 151 cm. There are three modes, 142 cm, 151 cm, and 155 cm, since each occurs three times and no other height occurs so often. The mean is 150 cm, to the nearest centimeter.

Spread. It is also important to have some information about the spread of a distribution. If two cities have an average temperature of 64° F, would you conclude that they are equally pleasant? You might, if that is the only information given. What if we also tell you that one city's temperature ranges in a typical year from $-15°$ F to 80° F? This last information uses the *range*—the difference between the highest and lowest scores. It is a common way to give information about spread. The box-and-whiskers plot showed the range, and also gave information about the *middle half* of the distribution—another useful measure of spread.

Predicting from Data

One type of higher-level problem involves predicting from data. For example, it is common folk wisdom among baseball fans that players who hit many home runs also strike out many times. Is the reverse also true—if a player strikes out many times, does he/she also tend to hit many home runs? If so, can we use the number of strikeouts a player makes in a season to predict the number of home runs he/she should hit? This can be illustrated by using data for the Detroit Tigers in 1968 —the last year they won the American League pennant. At that time, *all* players in the lineup, including pitchers, batted. A scatter plot for the team, showing strikeouts against home runs for each player, follows. Each dot represents one player, except the dots at 0 and 2 strikeouts, each of which represents two players. From looking at the scatter plot, it is clear that there is a tendency for those players with more strikeouts to hit more home runs. How can we use this information to predict, for example, how many home runs should be hit by a player who struck out 60 times?

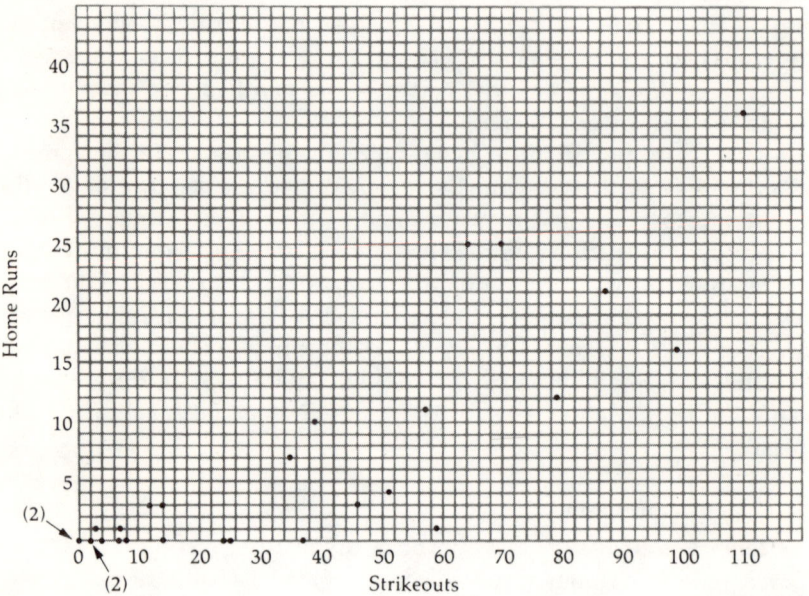

1968 DETROIT TIGERS

One way to do this is as follows: (1) count the number of points plotted—27 (one for each player); (2) divide the points approximately into thirds—which is easy to do here, 9 in each part; (3) find the median number of strikeouts and home runs in the left-most third—5 strikeouts and 0 home runs; (4) find the median number of strikeouts and home runs in the right-most third—70 strikeouts and 16 home runs; (5) draw a line through the two points found in steps 3 and 4. This is the line to use for predicting.

Find 60 strikeouts at the bottom of the graph. Come up to the line just drawn. Go to the left to read off the number of home runs. Sixty strikeouts should result in about 14 home runs.

The steps just described are illustrated in the graph that follows. Using the graph, how many home runs should a player hit if he/she struck out 100 times? If a player had 10 home runs, about how many times did he/she probably strike out?

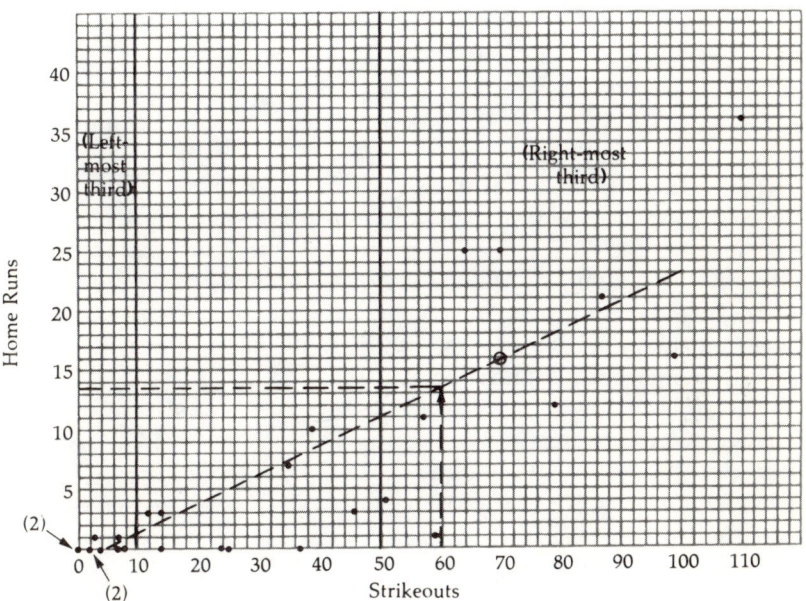

Making Inferences from Data

In the secondary school, data can be used for making inferences. Here are two examples where inferences can be used.

1. One hundred tosses of a coin result in 65 heads. Is it reasonable to suppose that the coin or the tossing process is biased in favor of heads?

2. A teacher has developed a new way to teach verbal problems to her algebra class. She uses the new method in three classes and the old method in two classes. The average score on a 20-point test is 14 for the new method and 11 for the old method. Is the difference large enough to be fairly sure that the difference is really the result of the new teaching method?

To answer questions like these means that one has to learn special techniques for measuring the importance of differences either between groups, or between what happened in an experiment (65 heads in the coin-tossing experiment) and what one would expect to happen (50 heads). Some of these techniques require a strong mathematical background and would be appropriate only in schools for junior or senior students with three or more years of college-preparatory mathematics. Other techniques involve simpler notions, such as counting, arranging scores in order, and simple mathematical operations. These techniques are grouped under the heading of *nonparametric statistics,* and could be taught to most high school students.

CONCLUSION

This article has looked at some ways in which people find and use data. Everyone deals with data nearly every day—collecting data, organizing data, interpreting data for oneself and for others, summarizing data with a few numbers, making predictions from data, and drawing inferences from data. Examples have been given, ranging from graphing at the elementary school level to working with more formal statistical methods in the senior high school.

The ability to work with data is a basic skill, one that at present is not adequately treated in most textbooks. Teachers need to provide experiences for students to work with data, and should introduce such work into the curriculum. Such experiences should not just be part of the mathematics curriculum, but they should also be incorporated into the science and social studies areas, where the data relate to real situations.

Data collection and analysis need not take large blocks of time from other study. Appropriate activities with data can provide students an opportunity to apply mathematical skills as they learn them, thus reinforcing regular mathematics instruction. Moreover, these activities are appealing and motivating to most students.

SEVEN

Problem Solving: Searching for Solutions
Mary Grace Kantowski

In one of his characteristically insightful moments, Mark Twain observed that everyone talks about the weather but no one does anything about it.

These days, everyone is talking about problem solving:

- The National Council of Teachers of Mathematics listed as its first item on the *Agenda for Action* (2) that "problem solving should be the focus of school mathematics of the 1980's."
- Presentations related to problem solving dominate national, regional, and local professional conferences.
- More and more pages of mathematics textbooks are being devoted to problem solving.
- Publications and projects related to problem solving are continually appearing.

Although everyone talks about problem solving, only the teacher in the mathematics classroom can do something about it. The teacher truly holds the key to the development of problem-solving ability in the student of the 80's. Curriculum developers and other mathematics educators offer direction and support, and textbook publishers provide a structure for instruction and a host of materials, but in the last analysis it is the teacher who meets the student day in and day out, who is charged with developing the problem-solving potential of each student to his or her capacity.

At this point many questions about problem solving and its role in the curriculum arise:

- What *is* problem solving?
- How significant is the role of the teacher?
- Are we calling for an end to the back-to-basics movement?
- Will the advent of the microcomputer have an effect on the problem-solving curriculum?

WHAT IS PROBLEM SOLVING?

There are many interpretations of the meaning of the word *problem*. To some, particularly those who work with children at the early elementary level, problems are verbal or "word" problems. To others, they are nonroutine situations, such as puzzles, logic problems or those requiring a bit of insight or creative thinking often found in "challenge" sections of textbooks. To still others, they are applications, or real-world problems. A problem can be "any of the above." What is a problem for one student may not be a problem for another—or even for the same student at a later time. For most students, textbook word problems are problems at first but they become exercises once students learn an algorithm or method to solve them. What makes a problem a problem is that, at least at first, there is some uncertainty about how to find a solution. Students have at their disposal, or can easily find, all the necessary facts but they are unsure of how to put them together to complete the task at hand.

Problem solving is the activity involved in trying to find a solution to a problem. This is a very important point and merits repeating: problem solving is the *activity* of searching for a solution. Problem solving is not necessarily finding a solution—although the ultimate purpose of instruction in problem solving is, to be sure, the development of the ability to solve a variety of problems. Before this goal is reached, however, the process of searching for solutions has a great deal of value in itself because it gives the student a sense of what works and what does not. Becoming an expert in problem solving is somewhat analogous to becoming an expert in gymnastics. Before the routines are perfected there are many spills and bruises and a great deal of discouragement, but in attempting new moves the gymnast develops a "feel" for what can be done.

In many aspects of mathematics a desired outcome is the knowledge of facts (such as the multiplication tables or common sets of Pythagorean triples) or skill in the use of an algorithm (such as the ability to apply the formula for the area of a triangle). In such cases the *answer* or the *product* is important. If problem solving is viewed as the activity involved in finding a solution, the solution itself no longer becomes all-important; the steps taken to find the solution take on greater significance. One important objective of instruction in problem solving is to help students develop the ability to approach problems, to learn to bring to bear everything they know that might be relevant to a particular problem, and to develop confidence in their ability to deal with unfamiliar problems. The activity or *process* of problem solving rather than the answer or *product* is what problem solving is all about.

This change of focus from the solution of a problem to the activity of problem solving has many implications for instruction. Much more time will have to be spent on *how* the solution was found, on how, for example, the decision to use a particular algorithm was made. Some techniques for dealing with novel situations are needed. Instruction for problem solving should include some rules of thumb (also known as heuristics) that will help students find ways to put together all they know to arrive at a solution.

George Polya (3) suggests a simple model for what goes on during problem solving that can serve as the framework for a model for instruction. According to Polya there are, or should be, four phases in the solution of a problem: understanding the problem, creating a plan, carrying out the plan, and looking back. When good problem solvers are studied in the process of solving problems, at least the first three phases are usually observed. Polya suggests that at each phase the problem solver ask himself or herself several questions. These questions can serve as an instructional tool for introducing the techniques of problem solving. During the introduction of problem solving, it is crucial to spend a good deal of time in the first two phases—understanding the problem and creating a plan. Because they have spent so much time in mathematics simply applying algorithms, many students skip these two phases and immediately jump into phase three of the model by trying to apply some formula or operation to the quantities involved. Questions based on Polya's phases serve as slow-down mechanisms that force students to *think* about the problem a bit before making a hasty decision to *do* something. Such questions as

> Would a diagram or figure help make the problem clearer? or
> How is what I'm given related to what I'm looking for?

help students focus on trying to understand the problem, while such questions as

> Have I seen a problem like this one before . . . how did I attack it?
> Could I organize my data into a table? or
> Is it possible to establish a pattern and generate a solution that way?

help students try to come up with a plan that might be useful in a given case.

A caution is in order here: Instruction in such techniques will be fruitless if students are not rewarded for using them. Talking about the importance of the activity of problem solving will fall on deaf ears if students receive credit only for correct solutions. Grading procedures must include some rewards for good efforts—even for those that fall short of a successful solution. That success will come in time.

HOW SIGNIFICANT IS THE ROLE OF THE TEACHER?

The importance of the teacher's role in the development of problem-solving ability cannot be overemphasized. Instruction in problem solving is, perhaps, more teacher-dependent than instruction in any other area of mathematics. The function of the teacher includes more than the obvious tasks of introducing techniques and strategies for use in problem solving and proposing sets of problems appropriate for particular students. The teacher has the capacity to pace the instruction according to students' needs—to give them help when they need it but to leave them to their own devices when they are engaging in fruitful activity. The teacher can answer questions (or perhaps not answer them), pose probing questions at opportune times, and present more challenging problems or different solutions to already solved problems when the occasion arises. The *Agenda for Action* exhorts teachers to "create classroom environments in which problem solving can flourish" (2). The enthusiasm with which a teacher reacts to novel solutions and intriguing problems may be intangible and not easily evaluated, but it is as important as any instructional technique. How many bright young (or older) scientists and mathematicians when asked to pinpoint the start of a love of mathematics that eventually led to a successful career reply, "It was Mr. Harvey back in the grades . . . he was so enthusiastic and gave me a curiosity about mathematics." or "Mrs. Travis who was the moderator of the Math Club made me experience the excitement of an intellectual challenge."

Because students view the same problem in different ways, one of the teacher's most important roles is to stimulate thought and to encourage a variety of ways to solve the same problem. The problem which follows is an interesting example. It is challenging enough for a mathematician or a college student who has not seen it before, yet simple enough to make it an excellent vehicle for discussion or group problem solving at the middle school, junior high or secondary school level:

Problem: The integers greater than one are arranged in columns as follows:

	2	3	4	5	
9	8	7	6		
	10	11	12	13	
17	16	15	14		etc.

In which column will 1000 fall?

An effective teacher can use a problem such as this to great advantage by having students generate a number of different solution paths (although there is only one solution) and by having students create problems that are similar. Recently, seven different solution paths were generated in a group of 17 students, each working independently. All students noted that 1000 would have to fall in either the second or the fourth column, and although all planned to generate rows until they found a pattern that could be generalized (or ground out) to find the desired result, the patterns they developed were very different. Some of the techniques used to find the solution included the following:

1. Multiplying each element of the table by 25 and generating rows to 1000.

2. Generating several more rows and noting that odd multiples of 100 are in column 4 and even multiples of 100 are in column 2.

3. Noting that rows beginning with even numbers are ascending and those beginning with odd numbers are descending; finding two possibilities for rows containing 1000, then finding the row with the form $16n + 14$.

4. Noting that initial elements of the outside rows differ by 8 and that numbers of the form $13 + 8n$ ascend in column 4 and numbers of the form $17 + 8n$ descend in column 2; then finding the closest n (123) and the row containing 1000.

Studying the variety of solutions to a problem such as this not only gives students a look at alternate ways to solve a problem, it also gives them a wealth of information about relationships among numbers and classes of numbers that can be useful in future solutions. Presenting one brief solution to the problem with the "answer"—the fact that the number 1000 would occur in column 2—would deny students the opportunity of engaging in some fruitful problem-solving activity and finding techniques and results that could be useful later.

Finding a solution to a challenging problem such as this is not always easy—in fact it can be downright difficult. What often happens in class, however, is that a nice solution that may have taken minutes (or even hours) to find is shared with others with a little handwaving and a few marks of the chalk. Wide-eyed students stare in amazement at a colleague or a teacher thinking, "I could never have come up with that." The truth is that they probably could have—with guidance and practice. Students need to see that problem solving involves some uncertainty and that they may follow many plans into dead ends before they find the path that finally leads to a successful solution. In short, they need to see that it isn't as easy as it looks. Occasionally solving a problem for the first time with students and sharing insightful moments with them can constitute a very effective lesson.

ARE WE CALLING FOR AN END TO THE BACK-TO-BASICS MOVEMENT?

It may seem that the call for greater emphasis on problem solving is incompatible with the back-to-basics movement. On the contrary, problem solving *is* a basic skill. In its Position Paper on Basic Mathematical Skills (1), the National Council of Supervisors of Mathematics listed problem solving as the *first* basic skill area, noting that this area includes such activities as posing questions, applying the rules of logic, determining which facts are relevant to a problem, and scrutinizing tentative solutions. This position does not minimize the importance of *computational skill,* it simply emphasizes the fact that there are important basic skills other than proficiency in the ability to perform the four basic operations.

Every teacher has had students who know the facts but who have difficulty in solving nonroutine problems: in the elementary grades are students who know their tables and who do very well on their drill sheets, but who cannot solve word problems; in the secondary school are students who know how to work with algorithms in algebra but who have trouble in applying the algorithms in application problems; in geometry classes are many students who do well in the computational aspects of the subject but who cannot complete an original proof.

These all too common situations support the NCSM position that skills other than computational skill are necessary for effective problem solving. This is not to say that computational skill is not important. Computational skill is a necessary condition for successful problem solving. It is not, however, a sufficient condition; other skills are necessary, and comparable emphasis must be placed on these other skills in instruction.

WILL THE MICROCOMPUTER HAVE AN EFFECT ON THE PROBLEM-SOLVING CURRICULUM?

The advent of the microcomputer should have a significant effect on the problem-solving curriculum of the 80's. The graphics mode of the computer has the potential to help students understand a problem more clearly by providing excellent color diagrams and simulating motion in a way not possible on the printed page or even in other media. Students can observe a diagram being drawn or see particular structures being highlighted in color. Using the keyboard or the paddles, they can actually move objects while working toward finding a solution.

The availability of the calculator mode of the computer also makes it an invaluable tool to use in teaching for problem solving. The school curriculum does not yet reflect the advent of the now readily available microcomputer to handle complex algorithmic and iterative calculations. Problems that are conceptually quite simple but that were formerly unsuitable for widespread use in mathematics because of tedious and time-consuming calculations can now be included earlier in the curriculum. With the computer available to do these calculations, the student is free to engage in the real activity of problem solving—the understanding, the planning, and the looking back.

CONCLUSION

During the 70's teachers and researchers learned a great deal about the processes of problem solving and about instructional techniques that are effective in promoting successful problem solving. During the 80's it is up to mathematics teachers to apply some of these techniques in their own classroom situations. Depending on students' problem-solving abilities, a teacher might assume one of many roles: that of model, groping with ideas and running into dead ends before reaching a fruitful solution path; that of resource person or "crutch," encouraging or helping students who make their own successful starts; or simply that of problem provider and facilitator for students who have gained some proficiency in problem solving. Whatever the role, the teacher must be an integral and active part of students' problem-solving endeavors.

Problem solving ability develops slowly over a long period of time (4). This development is not spontaneous, however; it requires painstaking effort on the part of both teacher and student. It will be ensured by instruction that includes emphasis on techniques that can be useful in solving problems, and practice with carefully selected sets of problems.

REFERENCES

1. National Council of Supervisors of Mathematics. "NCSM Position Paper on Basic Mathematical Skills." Minneapolis: The Council, 1977.

2. National Council of Teachers of Mathematics. *An Agenda for Action: Recommendations for School Mathematics of the 1980's.* Reston, Va.: The Council, 1980.

3. Polya, G. *How to Solve It.* 2d ed. New York: Doubleday and Co., 1973.

4. Wilson, J. W. "Generality of Heuristics as an Instructional Variable." Doctoral dissertation, Stanford University, 1967.

EIGHT

Calculators in Schools: Thoughts and Suggestions
Sherilyn Seitz and Terry E. Parks

The advent of inexpensive hand-held calculators in the early 70's initiated the potential for striking changes in mathematics teaching. As teachers require mathematics homework today, they must consider the fact that the vast majority of their students will have a calculator to use in completing the assignment. During the late 70's student ownership and access to calculators grew dramatically. Surveys in one district (Shawnee Mission, Kansas) indicated that by 1981 essentially all students, K–12, would either personally own or have access to a calculator at home (see Table 1).

Calculator availability to students outside school mandates that teachers implement practices which reflect their own or their school's policies in dealing with this fact. Teachers (and schools) should first determine whether or not they endorse the use of calculators in the classroom. In either case, instructional plans and activities are affected. The authors of this article and their school district have chosen to use and encourage the use of calculators in schools. The thoughts and suggestions herein reflect this philosophical commitment. The authors further believe that not to use calculators in classrooms is to ignore their prevalence outside school and, in students' eyes, to demonstrate a lack of realism or even hypocrisy.

The major concerns of teachers who hesitate to use calculators in

TABLE 1
Survey of Calculator Availability to Shawnee Mission, Kansas, Students

	Year of Survey			
	1975	1977	1979	(Projection) 1981
Number of students surveyed	23,088	22,230	16,721	—
Total enrollment	40,648	38,822	36,816	—
Percentage of district enrollment surveyed	56.8	57.3	45.4	—
Number of students owning calculators	2,210	6,659	7,545	—
Number of students owning/having access to calculators	11,341	16,046	14,439	—
Percentage of students owning calculators	9.6	30.0	45.1	(60)
Percentage of students owning/having access to calculators	49.1	72.1	86.4	(100)

their classrooms usually focus on a fear that computational skill development will be retarded by the calculator. A review of the current literature (4) showed that skills were not lost when children were allowed to have access to calculators. No measurable negative effects were associated with the use of calculators for teaching mathematics. The researchers observed that teachers who had not used calculators in the classroom believed there would be detrimental factors, whereas teachers who had used calculators in the classroom found them to be a worthwhile tool.

One study (3) asked teachers if students should master the basic skills before using a calculator. It found that the 80 percent of the teachers who had not used the devices in classrooms said yes, while teachers who had used them were twice as likely to say no. Several studies (1) have found that calculators do, in fact, help develop mastery of the four operations of arithmetic. The implications are that if the entire school community became familiar with the calculator and its potential for instruction, interest in and motivation toward mathematics would increase. This observable increase in interest is supported by the classroom experience of teachers in the authors' school district.

HISTORICAL BACKGROUND

What is the historical context for the use of calculating tools? Historically, the Chinese have been respected for their business acumen. Part of this ability has been credited to their computational speed in the marketplace, which has been enhanced by the abacus. These nonelectronic hand-held calculators have been used for centuries, without evidence of a lessening of their users' computational abilities. The abacus, or counting frame, was used by the ancient Greeks and Romans as well as the Chinese. Throughout history a "reckoning board" has been used to increase accuracy.

A more recent nonelectronic hand-held calculator was a small board containing numbers which a stylus pulled into place to give a reading of the total. Another version was a small plastic device invented for use in shopping to add costs by clicking ones, tens, or hundreds in columns as the shopper placed items in the shopping basket. Each device had drawbacks, often greater than its advantages.

With the invention of the microchip the hand-held calculator went through a metamorphosis. Decreases in size meant greater convenience and, with mass production, decreases in cost placed it within the reach of virtually everyone.

CLASSROOM APPLICATIONS AND ACTIVITIES

Teachers know that hands-on learning has always been effective with, as well as appealing to, children. Most people enjoy doing something themselves rather than just watching or listening to another. The calculator lends itself beautifully to learning in an active mode.

Calculators can enhance the student's appreciation of his/her own role in the teaching/learning process (2). Exploration leads to discovery of relationships and is often a reinforcement of concepts the teacher has introduced. For example, suppose the teacher has just introduced the "fours" in the multiplication tables. On the calculator, the student can supplement this learning by first entering 4, then the multiplication sign, followed by 1, and then trying to answer the problem before pressing the equal sign. This process gives immediate reinforcement for a correct answer or indicates an error. An effective procedure that makes the drill more palatable is to have the student repeat the equation until a correct match is made and then continue with $4 \times 2 = \ldots, 4 \times 3 = \ldots, 4 \times 9 = \ldots$. Such an approach to practice requires a minimum of adult supervision.

After the teacher has worked with children in learning multiplication (or addition, subtraction, division), it is helpful to let them check their own papers. Calculators can be located at the teacher's desk and checked out to students when they bring a completed paper to the teacher. The child can then address the task of checking his/her own calculations. It is a quick way for students to locate mistakes in the computation. If the answer on the calculator does not match the answer on the child's paper, the response may sometimes be: "This calculator isn't working." This is a clue to the teacher that a student needs help.

The calculator frees the teacher from awaiting tedious computation when introducing such concepts as partial products or the inverse processes of multiplication and division. The ease of handling large numbers with a calculator opens doors for exploring concepts that otherwise must wait for complete learner mastery and proficiency in computation. Consumer mathematics applications are examples. As inflation squeezes each penny tighter and tighter, consumer topics have become more and more important to individuals, and more and more teachers at all levels are incorporating consumer mathematics applications into their lesson plans. Some examples of classroom activities simulating consumer uses of mathematics follow.

>Set up a classroom store and ask students to bring empty food containers with the price stamp still showing. Cans and boxes showing ounces, pounds, grams, or liters, and a price provide for the determination of unit price. Comparison values can then be shown. Make shopping lists from the items available in the store. Find total costs by using the calculator.

>For a seasonal activity using a calculator to solve consumer mathematics applications from a newspaper, plan a holiday menu by shopping the advertisements of local grocers. Discuss and apply the sales tax using the calculator. Use gift lists and catalogs as well as other advertising media to extend the process.

>Discount houses offer catalogs that provide interesting consumer mathematics calculator projects. Assign the task of furnishing a baby's room with bed, bassinet, playpen, etc. Let another student furnish a whole house, or just a workshop. Compute the totals using a calculator.

>Interdisciplinary bulletin boards can be coordinated with a calculator center. For example, place a state road map on the bulletin board. Outline a trip to take around the state, indicating the highways to follow and the sites to see. Students can extend the activity by computing the cost of the trip when given such information as gas mileage and fuel costs, (for example, a vehicle gets 20 miles/gallon at a cost of $1.29/gallon). Let students devise their own trip, providing them with 3" × 5" cards on which to write problems for other students to solve. Such activities relate geography,

history, and mapreading, and when students create their own problems, they apply written communication skills.

Career and vocational education topics can be addressed using the calculator as a tool in a simulation of a real-life situation. For example, create a restaurant setting in the classroom in which one student takes orders and figures the bill. The customer can also check the restaurant bill with a calculator.

Role playing with charge accounts is another way to heighten student awareness of consumer concepts through calculator application. Students can compute charges mentally, then check with a calculator. They can also examine the addition of interest and the effect of finance charges.

Career and investment-related topics can be extended to include the stock market, commodities markets, money values, and exchange rates. For example, compute stock values daily from the newspaper quotations. Make graphs and note trends.

Traditionally, most classroom mathematics activities have used ditto, worksheet, or workbook. Too often they lack variety and student interest. A calculator is a learning aid and one that students enjoy using. Why not permit calculators for the homework assignment? Check out calculators to students and ask them to do such things as determine the gas mileage for the family car, or shop with a parent at the grocery store and total the cost as items go into the shopping cart. Students can use real data and the calculator to perform such studies as comparing—the price per ounce or gram of meat (hamburger, steak, etc.) and snack foods (potato chips, candy bars, etc.); the cost per calorie (or other nutritional unit) of different commodities. These kinds of applications are open-ended and lifelike.

Another benefit of calculator use in the mathematics classroom is in the determination of patterns and relationships. For example, at early elementary levels, when students are working with multiplication they "discover" that the product of 4×6 is four more than the product of 4×5, etc. They can also discover exponential concepts and patterns of multiples using the calculator's repeat function.

In teaching estimation teachers need some way to keep students' attention focused on the techniques of estimation. One way to do this is to place problems on an overhead projector and ask students to write their estimate of the answer. After going through a series of problems and seeing the written estimates, place the problems on the screen again and ask students to use the calculator to find the exact answer. Then discuss ways to enhance estimation skills. (Also see "Computational Estimation Is a Basic Skill," which appears earlier in this book, for other suggestions.)

Many books with calculator games are available. One of the oldest games, and a favorite with students, involves crossword puzzles using equations with correct answers that when turned upside down form words from the inverted numbers. For example, numbers converting to letters are $1 = i, 3 = e, 4 = h, 5 = s, 6 = g, 7 = 1, 8 = b$, and $0 = o$. To form the words "shell oil," the equation would need to have the answer 71077345 when solved. The task of designing equations with upside-down answers that must also fit into crossword puzzles is a fascinating challenge to students.

All the ideas just suggested have assumed the availability of a basic four-function calculator. The selection of an appropriate calculator for a classroom depends on the applications planned for students. The authors have found the use of a four-function liquid crystal display device with a 2,000-hour battery life to be entirely satisfactory for most classroom uses K–12. Some higher-level mathematics classes need trigonometric functions and memories, but for most applications these features are not essential. The primary caution is to purchase calculators with a 2,000-hour battery life (or longer) using AA batteries. They may cost more initially, but less expensive machines with a shorter battery life will cost much more over time with the added inconvenience of having to change batteries frequently.

Numerous models of programmable calculators are now on the market. For mathematics—as well as science, social studies, business—these machines have value. Sometimes called "smart calculators or dumb microcumputers," they can perform complex statistical analyses in a few minutes. They can solve quadratic equations, multiple resistor networks, or inductor/capacitor frequency and impedance problems in seconds. They can be used to play games such as black jack, dice-toss, number guessing, moon landing, submarine, cannonade, biorhythm. Practical uses of the programmable calculator are many and varied. Some examples are computation of mortgage payments, bank statements, compound interest, and discount rates.

CONCLUSION

In conclusion, the following recommendations are based on the authors' own classroom experience:

- Use calculators in your classroom.
- Continue to use manipulative devices and concrete aids to present concepts. Calculators are not a substitute for these tools.
- Let students use calculators to solve word problems. The objective of word problems is problem solving and logical thinking, not drill in computation.
- Use as much "real" data (that is, the kinds of problems found outside textbooks) as possible. Word problems need not be limited to those in textbooks.
- Teach mathematics, not calculator technique. Calculators are so common that special lessons on their use should not be necessary. A few instructions will suffice. Remember that cal-

culators are simply a tool, a means to an end.
- Try some of the activities suggested in this article with your classes.
- Develop your own activities for problem solving using calculators.

REFERENCES

1. Suydam, Marilyn N. "The Use of Calculators in Pre-College Education: A State of the Art Review." Calculator Information Center, Columbus, Ohio, May 1979.

2. Wheatley, C. L. "Calculator Use and Problem-Solving Performance." *Journal for Research in Mathematics Education* 11, no. 5 (1980): 323–34.

3. Wheatley, Grayson A.; Shumway, Richard J.; Coburn, Terrence G.; Reys, Robert E.; Schoen, Harold L.; Wheatley, Charlotte L.; and White, Arthur L. "Calculators in the Classroom." *Arithmetic Teacher* (September 1979): 19–21.

4. Wyatt, J. Wendell; Rybolt, James F.; Reys, Robert E.; and Bestgen, Barbara J. "The Status of Hand-Held Calculator Use in School." *Phi Delta Kappan* (November 1979): 217–18.

NINE

Computers in School Mathematics
J. D. Gawronski

Computers and computer applications are everywhere these days —in schools, classrooms, fast food franchises, offices, banks, and on and on. The microprocessor technology has led to computer-monitored microwave ovens, home security systems, television recording systems; it is relied on to report sports events and election returns. Neither child nor adult can avoid the influence of the computer in today's world.

The technological age has introduced computer-related "hardware," "high" technology, and "software" into our lives and into our language. The easy access to high technology has come about because of dramatically declining prices for equipment, high reliability, and the fact that now nearly everyone knows someone who uses a computer. The computer is no longer a "magical," unfamiliar tool that is used only by researchers or scholars or scientists. It is a remarkably effective tool that is finding its way into home, school, and business because it is too useful to do without. The computer helps us do our jobs and even routine tasks more effectively and efficiently. More importantly, it gives us power never before available to solve complex problems.

All of this "revolutionary" change has particular import for the school curriculum. Computer literacy skills are basic skills that everyone needs or will need in order to function comfortably and effectively in the future. Knowledge of what a computer is, and of what it can and cannot do, is part of the minimum competencies that all students should be expected to attain.

Awareness of what a computer program is and of its relationship to a computer is fundamental knowledge that should be included in school mathematics programs. This awareness is not the same as learning to program—it is simply an acknowledgment, a realization, an understanding that it is the program, the software, that directs a computer in what it does. Change the program and the computer's operation changes. Different people using the same computer program will obtain the same results, but one individual using different programs will obtain a different result for each program. This understanding is prerequisite to acquiring computer programming ability, and it is most certainly a computer literacy objective for all students.

In addition, the knowledge of how to access, store, retrieve, and use information in a computer environment becomes critical. These information-processing skills must also be included in the basic school program for all students. Increasingly, more job and career opportunities are dependent on computer literacy and information-processing skills.

Professional associations of mathematics teachers have responded to the perceived need for computer literacy. The National Council of Teachers of Mathematics, in its *Agenda for Action: Recommendations for School Mathematics of the 1980's,* stated explicitly that "mathematics programs must take full advantage of the power of calculators and computers at all grade levels." Although many individuals and groups strongly advocate computer literacy, there is no clear consensus on the specific skills a computer-literate person should possess. Because of a lack of consensus at this early stage, mathematics teachers have often defined operationally in their course design their meaning of computer literacy. As a result, certain themes are becoming common and widely accepted. These themes include the following:

1. What a computer can and cannot do
2. What a program can and cannot do
3. How to program

Students need to learn to evaluate the advantages and disadvantages of using computers to perform particular tasks or to make selective application. Not all problems are best solved by a computer. But students can appreciate that many problems, such as extensive data analysis, simulations, analysis of alternative models, can be solved more effectively, more quickly, and in some cases exclusively, by the considerable power of a computer program.

To learn what a computer can and cannot do requires some hands-on experience. And the appreciation of computer functions and limitations includes a knowledge of computer uses in everyday life.

Computer literacy also includes knowing what a "program" is and what it is capable of accomplishing. Computers must be given directions. This may seem obvious, but there is widespread misunderstanding of this fact among the general public. When students use, test, and evaluate programs, they learn how the set of directions called the "program" determines the way in which the computer solves a problem or plays a game. Young people are especially attracted to computer games, and when they ask such questions as "Why does it work that way?" or "How does the computer know what to do?" they are on the way to understanding what a program does. Often their desire to make the program do something differently leads to experimentation which aids them in acquiring programming skills.

Finally, a computer-literate person should know how to write an original program. The computer can enhance problem solving, but programming itself requires problem-solving ability and acumen. Programming can also be an aid to an understanding of a mathematical concept or an algorithmic technique.

The pervasiveness of computers raises crucial questions about the mathematics curriculum. How, when, and what to add to the curriculum, as well as what to remove from the curriculum, become significant issues that must be addressed if students are to be prepared for this technological age.

Mathematics teachers have begun to confront these concerns since in many school systems they were the first to introduce computers in the classroom. In the 60's and early 70's these efforts were relatively meager because of the high cost of both computer hardware and telephone lines to support time-sharing systems. Recently, however, the microprocessor has made low-cost computer power available and within the reach of many individual schools and districts. Thus, the potential for curriculum change is increased for two reasons. First, the low cost of the microcomputer makes it attractive to both school program directors and to individual consumers. Second, the more computers are bought for personal or business use, the more the public will expect the schools to be teaching about computers and their uses. Thus, the low cost of equipment which permits easy access, and changing cultural expectations become forces moving the school mathematics program toward inclusion of computer literacy and information-processing skills.

Since schools must take an active role in preparing students to live, work, and play in a world society in which computers are performing more and more functions, there must be a profound rethinking of what is basic now and what will be basic in the future. For many years, facility with computational skills and algorithms has been considered an essential, basic, minimum skill. But is this still the case? The importance of addition, subtraction, multiplication, and division skills should not of course be underestimated. These skills, however, are only the tools and techniques that help us solve problems. The knowledge of when to use or apply these tools is the crucial skill. Once we know that addition or multiplication can solve a problem, the calculator and/or computer can assist and carry out the computation.

Consider the present role of the square root algorithm. The concept of square root is certainly an important and useful one. However, the critically important knowledge is knowing when it is necessary to calculate a square root to solve a problem. Once this decision is made, the problem solver reaches for a calculator or includes a statement in a computer to determine the square root. A few years ago the problem solver would have reached for a table of square roots or a slide rule. For many years now, problem solvers who use the square root algorithm to calculate a square root have been rare.

With the widespread use of calculators and computers, it is no longer necessary to develop the skill of completing long and complicated calculations quickly and correctly. Rather, the skill of estimation, the ability to recognize the reasonableness of a result, becomes basic. For example, the cash register keyboards in some fast food franchises contain pictures of the food rather than numbers to indicate dollars and cents. When a customer purchases a hamburger, french fries, and a malt, the clerk hits the keys with the pictures of these items, and then the register displays the total price. This is not science fiction—this is today. It is probably more time-efficient to hit one key with a picture of a hamburger than to hit two or three keys to indicate price. When prices or menus change, extensive staff retraining to learn new prices or to match prices to new items is not necessary. A few simple programming changes in the computer can make the changes quickly.

But what skill do both the customer and the clerk in this scenario need? They need to be able to estimate the answer to recognize that the total amount shown is reasonable. If the total indicates $.17, the clerk should certainly be quick to recognize a mistake in the customer's favor. And if the total indicates $17.00, the customer should recognize a mistake in the store's favor. In such a work situation, the clerk is no longer expected to add and subtract quickly and accurately to determine totals and change. Once the clerk enters the information, the computer-driven cash register completes these calculations. The clerk as well as the customer, however, should be able to judge or estimate if the result shown is a reasonable one.

The computer should serve to extend, to augment, to facilitate curriculum development and curriculum change. School mathematics programs must not be confined to extensive and repetitive practice of what will become archaic computational skills. The focus should be on acquisition of concepts of addition, subtraction, multiplication, and division, with multiple opportunities to apply these concepts. The ability to add columns of four-digit numbers is not a highly prized skill in today's real world. Nearly everyone who performs such tasks uses a calculator or a computer program to perform them. The more highly prized skill is knowing *when* to add to solve the particular problem under study.

What does this shift mean for the school mathematics program? It means that the focus can move from practicing the "tool" to learning how, when, where, and why to use the "tool." Relevant skills become (1) an understanding of the operation or concept, (2) an ability to recognize when to apply the operation appropriately, (3) an ability to use a calculator or computer appropriately. The computer can do the rote, tedious part of "crunching the numbers," and the teacher and the student are free to explore, apply, use, and practice mathematical skills in meaningful ways.

In addition to such curriculum content shift or change, information-processing and computer-literacy objectives and activities should become prominent in the K–12 program.

There is a need for better educational software and for more programs that not only enhance problem solving but take full advantage of the unique capabilities of the computer—capabilities unavailable in other instructional media. But existing software can serve as an introduction. Many students have learned to program by working with a "canned" program and by experimenting with changes to make it do something else. Such changes of existing programs can lead to a study of what program statements do and eventually to the development of an original program. This strategy is similar to the way in which a child learns to speak the native language. The native language is an existing structure which surrounds the child in his or her environment. The child learns to imitate in the language, to explore it, to create original phrases, and eventually to put the language together in an individual style. In similar ways, the student can create original programs by exploring and modifying previously created software.

Because of societal demands and the rapid development of new computer technology, many of these curricular changes have begun. The burden for systematic change, however, rests with schools and mathematics teachers. Some of the groups that must work together to bring about a responsive school program include professional teacher organizations, curriculum developers, community groups, and school administrators. The exponential growth in computer capability and access makes such change a critical test of effective schooling.

In summary, the impact of computers on school mathematics programs is beginning to be felt, and this impact will increase dramatically. The use of computers in the classroom affects the curriculum in at least three ways:

1. The mathematics curriculum is expanded to include new goals and objectives such as computer literacy and information-processing skills.
2. Traditional mathematics teaching approaches such as "show and drill," repetitive practice on long computational algorithms are minimized.
3. Mathematics objectives in estimation and problem solving receive increased priority, and learning activities related to developing these abilities receive increased attention and time.

Throughout the 80's, these three effects are certain to continue and to increase in magnitude. Consequently, they will be among the major determinants of the future school mathematics curriculum.

TEN

Women and Mathematics: Is There a Problem?
Mary Schatz Koehler and Elizabeth Fennema

In the past decade, there has been a great deal of concern as well as much research and scholarly writing at all levels about sex-related differences in mathematics. Is there really a problem? Are females being short-changed in mathematics education? Does it make any difference if females do not learn mathematics? In order to get a clearer focus on the future and to determine what should be done, let us review the past status of women and mathematics, and also consider the current situation.

PAST STATUS

As the lack of prominent female mathematicians in the course of history indicates, women have faced some hard times in mathematics. In earlier centuries it was thought that "women were unsuited for the study of this subject [mathematics] because their heads were smaller than men's, their nervous systems too delicate, or their intellectual capabilities not sufficient to the task" (4, p. 262). Gauss, an eminent German mathematician of the eighteenth century, stated that "when a person of the sex, which, according to our customs and prejudices, must encounter infinitely more difficulties than men to familiarize herself with these thorny researches, succeeds nonetheless in surmounting these obstacles and penetrating the most obscure parts of them, then without doubt, she must have the noblest courage, quite extraordinary talents, and a superior genius."

Unfortunately, such views are not limited to previous centuries. Only a decade ago, Aiken said "sex differences in mathematical abilities are, of course, present at the kindergarten level and undoubtedly earlier" (1, p. 203). And, although well refuted, the statement "we favor the hypothesis that sex differences in achievement in and attitude toward mathematics result from superior male mathematical ability" was made in 1980 (3, p. 1264).

Undoubtedly, we are all familiar with the grim statistics showing how few women throughout history have participated in science or mathematics-related careers. For example, in the 50's women earned only 6.7 percent of all science and engineering doctorates awarded, and in the 60's only 7.9 percent of these degrees. The 70's saw an increase to 14.9 percent, which is still far below women's representation in the population (20).

Sells (19) has called mathematics the critical filter into many college majors and careers. In looking at the mathematics course background of students admitted to the University of California, Berkeley, in 1972, she found that 57 percent of the men and only 8 percent of the women had sufficient high school mathematics to qualify for calculus. Since many college majors require the calculus sequence, students who enter college underprepared to take calculus are severely limiting their options.

Women have also been underrepresented in many fields requiring a vocational or technical school diploma. For example, in 1978 only .2 percent of all electricians were women (8). Mathematics serves as a critical filter into many of these fields as well. Here are just a few which either require or recommend more than high school algebra: animal technician, graphic artist, dental hygienist, occupational or physical therapist, chemical technician, communications worker, land surveyor, and mechanical draftsperson (12, pp. 78–79).

CURRENT SITUATION

We have just seen that attitudes toward women and mathematics have been substantially less than favorable, and that women have been severely underrepresented in many careers involving mathematics. We will now assess the present-day situation, by considering three components: mathematics achievement data, high school course enrollment data, and career data.

Achievement Data

In an extensive review of research, Fennema found "that there are no consistent significant differences in the learning of mathematics by boys and girls in the early elementary years" (9, p. 128). Research involving older students led to the conclusion that "in overall performance on tests measuring mathematics learning, there are no significant differences that *consistently* appear between the learning of boys and girls in the fourth to ninth grade. There appears to be a trend, however, that if a difference does exist, girls tend to perform better in tests of mathematics computation and boys tend to perform better in tests of mathematical reasoning" (p. 135). Partly because of the paucity of data, and partly because of confounding factors, no conclusions could be reached about sex differences in high school mathematics.

The mathematics assessment of the second National Assessment of Educational Progress (NAEP II) provides more recent information. In 1978 this study assessed the learning of a random sample of 70,000 9-, 13-, and 17-year-olds in five content areas (number and numeration, variables and relations, geometry, measurement, and other topics) at four cognitive levels (knowledge, skill, understanding, and application). In reviewing the results, Fennema and Carpenter conclude that "the assessment results indicate that on a nationwide basis, there is little difference between males and females in overall mathematics achievement at ages 9 and 13. At age 17, however, females are not achieving at the same level in mathematics as are males. Even when females and males reported they had been enrolled in the same mathematics courses, males' performance was higher than that of females, and the differences were greatest on the more complex tasks" (13, p. 6).

Another source of current information on mathematics achievement is the data from the mathematics portion of the Scholastic Aptitude Test (SAT). This test is taken by approximately one million high school seniors annually, students who represent about one-third of all high school seniors and about two-thirds of all seniors who go directly to college. The data that follow, then, are not meant to represent all high school seniors, but rather a self-selected sample, who, although they represent a wide range of abilities are "more apt in comparison to all high school seniors" (7, p. 4).

Table 1 gives the data from the mathematics portion of the SAT for the last eight years. Each year males in this group outperformed females. Although the scores for both males and females have decreased, the difference between female and male scores has widened over the years.

TABLE 1
SAT—Mathematics Mean Scores, 1972–1979

	Male	Female	Difference
1972	505	461	44
1973	502	460	42
1974	501	459	42
1975	495	449	46
1976	497	446	51
1977	497	445	52
1978	494	444	50
1979	493	443	50

Source: College Entrance Examination Board. *College-Bound Seniors: National Report, 1971–1979.* (Princeton, N.J.: Educational Testing Service, 1979.)

After examining this achievement data, one could reasonably conclude that while differences in favor of males are not manifest before high school, in the later high school years differences in favor of males do appear. This is not to say that all males perform better than all females on all measures of mathematics achievement, but simply that males as a group tend to outperform females as a group in the later high school years, on some measures of mathematics achievement.

Enrollment Data

We have mentioned previously that mathematics is a critical filter to many careers, and to undergraduate and technical school majors. Enrolling in high school mathematics courses is one way for students to keep their options open. Let us now consider the question of whether there are differences in the pattern of enrollment between males and females. NAEP II and the SAT serve as sources of enrollment data.

Table 2 gives the percentage of 17-year-old students in the NAEP II testing who reported that they had been enrolled for at least one half-year in a particular mathematics course. This table shows that in the beginning high school courses there is very little difference in the course-taking patterns of males and females. However, a significantly

TABLE 2
Mathematics Course Background

Course	Percentage of 17-Year-Olds Having Taken at Least 1/2 Year	
	Females	Males
General or Business Math	47	44
Pre-Algebra	45	46
Algebra I	74	71
Geometry	51	52
Algebra II	36	38
Trigonometry	11	15
Pre-Calculus/Calculus	3	5

Source: Fennema, E., and Carpenter, T. "Sex-Related Differences in Mathematics: Results from National Assessment." *Mathematics Teacher,* in press.

higher percentage of males than females were enrolled in the more advanced classes of trigonometry and pre-calculus/calculus (13). It should be kept in mind that most subjects in this sample were eleventh grade students. The differences may have been greater if twelfth grade students had been assessed.

Although representing a self-selected sample, data gathered from those students taking the SAT give more course-enrollment information. Table 3 shows that higher percentages of females than males have

TABLE 3
Years of Mathematics Study as Reported by SAT Participants

	2 years		3 years		4 years		5+ years	
	M	F	M	F	M	F	M	F
71–72	11[a]	19	29	40	50	33	9	4
72–73	10	20	28	40	51	33	9	4
73–74	11	21	29	40	50	33	9	4
74–75	10	20	28	39	50	34	9	4
75–76	10	20	27	37	51	35	10	5
76–77	10	19	26	36	50	35	11	5
77–78	10	18	25	35	51	37	12	6
78–79	9	17	24	34	53	39	12	6

Source: College Entrance Examination Board. *College-Bound Seniors: National Report, 1971–1979.* (Princeton, N.J.: Educational Testing Service, 1979.)
[a] All numbers reported are percentages.

taken only two or three years of high school mathematics, while higher percentages of males than females have taken four or five years of mathematics. (A high school student is considered to have taken five years of mathematics when she/he has taken a college-level course such as calculus.) In fact, for the past eight years, over 50 percent of the males who took the SAT reported that they had taken four years of mathematics in high school. For this same period, the percentage of females taking four years of mathematics has increased at a more rapid pace than the percentage of males, but female enrollment still lags behind that of males.

It is evident then, that females are not enrolling in advanced high school mathematics courses in the same proportion as males. This lack of participation in high school mathematics can eventually lead to female underrepresentation in many careers. As Fennema points out, "Although only symptomatic of the effects of many variables, electing not to study mathematics in high school beyond minimal or college requirements is the cause of many females' nonparticipation in mathematics-related occupations" (10, p. 7). We will now look at some career data to see where females are employed.

Career Data

Not only are females, to a large extent, still employed in traditionally female fields, but female students are still making traditional career choices. For example, some professions with fewer than 10 percent female members are engineering, law, medicine, and dentistry. On the other hand, females comprise 98 percent of dental assistants and secretaries (8).

Students taking the SAT provide us with information regarding their intended career areas. In 1979, choices that were predominantly male (greater than 70 percent) were military science, engineering, architecture, geography, physical science, and forestry. In the same year, choices that were predominantly female (greater than 70 percent) were home economics, library science, foreign languages, psychology, education, art, theater arts, English/literature, and health and medicine (7).

Thus it is evident that women are still not found in many technological and scientific careers, nor are they planning to enter these fields. Since many of these careers are financially rewarding, women are being short-changed by not being encouraged to enter them. However, the nation as a whole is also being short-changed by not receiving the contributions of many talented women.

Notwithstanding the progress that has been made, then, we see that sex-related differences in mathematics achievement, enrollment, and career choice still exist. Females are still receiving a poorer mathematics education than are males.

VARIABLES RELATED TO MATHEMATICS EDUCATION INEQUITIES

There are, of course, many variables that can influence an individual's mathematics learning. Some have been shown to have more impact than others. We will consider three categories of variables: attitudes, influences of significant others, and influences of schools.

Attitudes

Although attitudes are not developed independently of achievement, they are highly related to the learning of mathematics. Three specific attitudes help us understand why sex-related differences in mathematics exist: (1) confidence in one's ability to learn and perform well in mathematics, (2) perceived usefulness of mathematics, and (3) perception of mathematics as a male domain.

Confidence. Confidence, or the belief that one can do well in mathematics, is positively correlated with mathematics achievement. In fact, it is almost as highly correlated as are the intellectual variables of spatial visualization and verbal skills. Girls report less confidence than boys in their ability to do mathematics even when they are in fact achieving as well as boys. This lowered confidence is evident as early as sixth grade and perhaps earlier (15). The importance of confidence to one's learning of mathematics can be summed up by saying that "this finding of less confidence by females influences how hard they study, how much they learn, and their willingness to elect mathematics courses" (11, p. 9).

An examination of the reasons females give for the causes of their successes or failures shows one way that they demonstrate their lack of confidence. Females, more than males, attribute their successes to luck or to some environmental influence, while males, more than females, attribute their successes to their own ability. Males tend to believe that they are in control of the situation. Because of this control, they expect repeated success. Females perceive that the reason they succeed is not within their control, and consequently tend to believe that success will not repeat itself. The trend reverses in discussions of failure. Females tend to blame failure more on their own lack of ability, while males blame failure more on luck or on the environment. Since females often do not believe in their own ability, they tend to expect failure and often have little or no confidence in their ability to perform well in the future.

Perceived Usefulness of Mathematics. Whether or not a student believes mathematics will be of personal value is another attitude crucial to the learning of mathematics. Students who perceive mathematics as useful in either educational or career plans will be motivated to put more time and effort into studying, and to elect more mathematics courses. On the other hand, students who do not perceive it as relevant to their goals will not be likely to invest the time and energy necessary to obtain a solid understanding of mathematics.

Starting in junior high school, males perceive mathematics as useful to them to a much greater degree than do females. Males, much more than females, would tend to agree with the following items from the Fennema-Sherman Mathematics Usefulness Scale (14):*

1. I'll need mathematics for my future work.
2. I study mathematics because I know how useful it is.
3. Knowing mathematics will help me earn a living.
4. Mathematics is a worthwhile and necessary subject.
5. I'll need a firm mastery of mathematics for my future work.
6. I will use mathematics in many ways as an adult.

The fact that males tend to perceive mathematics as useful to a greater degree than do females, helps explain why they elect more mathematics courses than do females, and why they are consequently more often employed in fields that use mathematics than are females.

Mathematics as a Male Domain. Repeatedly in our society, mathematics and mathematics-related work are seen as masculine. Unfortunately, this perception of mathematics as a male domain is not some archaic view held only by "older generation" men and women. Osen notes that "many women in our present culture value mathematical ignorance as if it were a social grace" (18, p. ix).

One of the most consistent findings from research on sex-related differences in mathematics is that males, more than females, stereotype mathematics as a male domain. For example, starting as young as sixth grade, boys agree more than do females with statements such as

1. It's hard to believe a female could be a genius in mathematics.

---

*The "Fennema-Sherman Mathematics Attitude Scales, JSAS *Catalog of Selected Documents in Psychology* 6, no. 1 (1976): 31 (ms. no. 1225), are available for $5.00 from the American Psychological Association, 1200 17th St., NW, Washington, DC 20036.

2. When a woman has to solve a math problem, it is feminine to ask a man for help.
3. I would have more faith in the answer for a math problem solved by a man than a woman.
4. Girls who enjoy studying mathematics are a bit peculiar.
5. Mathematics is for men; arithmetic is for women.
6. I would expect a woman mathematician to be a masculine type of person. (14)

Considering that at the junior high and high school level there are more male than female mathematics teachers, and realizing further that many mathematics-related occupations are predominantly male bastions, it is not surprising that students view mathematics as a male domain. Unfortunately, this stereotyping occurs at a crucial time. As Fennema explains, it is probably "more than just coincidence that at adolescence, when girls are becoming increasingly aware of their sex role, sex-related differences in mathematics learning appear" (11, p. 11).

Individuals do those things they see as appropriate for their sex. If a girl perceives an activity as feminine, she will be more apt to participate in it. The same influence works on boys. If a boy perceives an activity as appropriate for males, then he will feel more comfortable performing it. Not only do individuals tend to select activities perceived as appropriate for their sex, they fear sanctions from others if they perform opposite sex-stereotyped activities. In relation to mathematics, females may fear social rejection if they excel in mathematics, while males will be pressured into doing well.

Influences of Significant Others

A second set of factors that affect sex-related differences in mathematics is the influence of teachers, counselors, peers, and parents on students.

Teachers. Although many teachers feel helpless in influencing students, we firmly believe that teachers are the most important influence in students' learning of mathematics. Students often point to a single teacher as the cause of either their liking and electing mathematics courses, or their disliking and avoiding them. Research based on observations of mathematics classrooms has found that teachers treat males and females differently in both subtle and not-so-subtle ways. Teachers interact more with males than they do with females. They pay more attention to high-achieving males than to high-achieving females or to any other group. They expect boys to do better on higher-cognitive-level tasks, and thus encourage and call on them more frequently when the mathematics is of a higher cognitive level. They encourage boys to behave independently and to persist in finding solutions to difficult problems, but they encourage girls to be dependent. Teachers can influence girls. When teachers hold the same high expectations for girls as they hold for boys, girls perform as well as boys (5).

One result of this differential teacher treatment of boys and girls is that boys become more autonomous learners of mathematics than do girls. That is, boys become "thinkers who are independent problem solvers and who do well in high-level cognitive tasks" (11, p. 13). Girls, on the other hand, learn to be dependent and "helpless" with respect to problem solving.

Peers. One has only to watch adolescent girls and boys to confirm the idea that peer influence is important. Extremely interested in their peers' opinions, adolescents often tailor their behavior to harmonize with their perception of their peers' expectations. Since boys, much more than girls, stereotype mathematics as a male domain, they no doubt send many subtle, and not-so-subtle, messages that girls who achieve in mathematics are somewhat less feminine.

Counselors. During the high school years when students seek advice concerning elective courses, counselors can be very influential. Unfortunately, counselors often uphold and reinforce the stereotype of mathematics as a male domain, and tend to view mathematics as less important for females. Not only do counselors fail to encourage females to elect mathematics courses, they often discourage them from electing these courses. Counselors may not be aware that by not actively encouraging them to take mathematics, they are effectively closing many educational and career options to females.

Parents. In many ways, the attitudes and stereotypes that parents hold are passed on to their children. A student who receives parental support and encouragement to work hard in mathematics, and who receives parental approval and praise for excelling in mathematics, is much more likely to persist in the subject than one who does not. Also, a student whose parents view mathematics as useful and who encourage her or him to elect more advanced mathematics courses in later high school and postsecondary years will be more likely to do so. Parents are more likely to discuss course decisions and career plans with their sons than with their daughters, and are more supportive of their sons' mathematical interests. They also hold lower educational aspirations for their daughters than for their sons (16).

More so than fathers, mothers often have inadequate mathematics backgrounds and hold negative attitudes toward mathematics. Since daughters often look to the mother as a role model, the mother's feelings about mathematics can be critical. A mother who lacks mathematics skills is likely to accept her daughter's poor mathematics grades as inevitable. As one seventh grader told the authors, when explaining a girl's negative attitude toward mathematics, "She has it because her mother pretty much has it and . . . it's just been sort of passed down. She just has caught it from her mother."

Overall School Influences

The influences of significant others, especially those of teachers and counselors, should not be read as *"all* teachers do such-and-such" or *"all* counselors behave thus-and-so." There are as many individual differences among teachers and counselors as there are among females and males. Many schools do have high percentages of females not only enrolled in advanced mathematics, but also performing well in such courses (6). In these schools, the teachers are a powerful influence in persuading the females to enroll in higher-level mathematics courses. The teachers assumed the roles of "trusted older friend, respected mentor in their field of interest, and that of informed and aggressive counselor" (6, p. 156).

WHAT CAN BE DONE?

So far, we have presented a problem (that of sex-related differences in achievement and enrollment in mathematics) and we have discussed some factors influencing or causing that problem. Now, however, we need to look at how to go about implementing a solution to that problem. We have several suggestions, and we have focused them on teachers and on schools.

Teachers

Teachers need to be more aware of their impact on students. They need to become sex-blind with respect to their teaching of mathematics —that is, they need to treat males and females the same. They need to encourage both males and females to keep their educational and career options open by electing more mathematics classes. They need to help all students develop feelings of confidence about their ability to do mathematics. Teachers should not attempt to bolster confidence by lowering their expectations. They should hold high expectations for both males and females. Teachers should continue to monitor the textbooks and media aids used to be sure that they do not perpetuate the stereotype or myth that mathematics is a male domain.

To further these ends, teachers can ask themselves several questions such as the following:

1. Are there sex-related differences in mathematics learning in my classroom? Are boys more apt to be better problem solvers than are girls?
2. What are my students' attitudes toward mathematics? Do the girls feel differently about themselves as learners of mathematics?
3. Do I treat girls and boys differently in my classroom? Do I call on boys more than I call on girls?
4. Do I have different expectations in mathematics for girls and boys? How do my expectations affect my interactions with girls and boys?
5. Do I give information to both boys and girls about the usefulness of mathematics?
6. Do I use any instructional materials that are sexist?

Schools

Schools, too, can do much to help alleviate the problem of women and mathematics. Most importantly, they can reach four important groups—teachers, counselors, parents, and students—and disseminate information to them regarding the usefulness of mathematics. The fact that mathematics opens the door to countless educational and career opportunities cannot be stated often enough. Even though most colleges require only one or two years of high school mathematics for admission, most mathematics majors require three or four years of high school mathematics. For the non-college-bound student, mathematics is also important. Our world is becoming exceedingly technological and scientifically complex; therefore an understanding of mathematics is simply a basic survival skill.

Schools also need to guide these four groups in working to eliminate the stereotype of mathematics as a male domain. Boys and girls need to be shown that both sexes are quite capable of, and should strive for, excellence in mathematics.

Schools, too, can ask themselves a series of questions.

1. Are there sex-related differences in enrollment patterns when mathematics is elective? If so, what should be done about them?

2. Are gifted females, as well as gifted males, given extra support and encouragement in mathematics?
3. Are females, as well as males, specifically encouraged to actively participate in mathematics-related activities such as computer clubs?
4. Are counselors consistently informed about the importance of mathematics for both girls and boys?
5. What is this school actively doing to inform parents about the usefulness of mathematics for girls and boys?

There are several intervention programs that schools can implement to more forcefully combat the problem of women and mathematics. One such program is "Multiplying Options and Subtracting Bias,"* a series of four videotapes and workshops specifically designed for teachers, counselors, parents, and students. It provides accurate information about women and mathematics and addresses such topics as the usefulness of mathematics, the stereotyping of mathematics as a male domain, confidence in learning mathematics, and differential treatment of males and females as learners of mathematics.

Teachers and schools may well come up with solutions suited to their own situation. It is critical, however, that all groups recognize and work on the problem of inequitable mathematics education for females. All should join the National Council of Teachers of Mathematics in their commitment "to the principle that girls and women should be full participants in all aspects of mathematics education. Both simple justice and future economic productivity require that we do so without further delay" (17).

REFERENCES

1. Aiken, L. R., Jr. "Intellective Variables and Mathematics Achievement: Directions for Research." *Journal of School Psychology* 9 (1971): 201–212.

2. ———. "Some Speculations and Findings Concerning Sex Differences in Mathematical Abilities and Attitudes." In *Mathematics Learning: What Research Says About Sex Differences*, edited by E. Fennema. Columbus, Ohio: ERIC Center for Science, Mathematics, and Environmental Education, 1975.

3. Benbow, C. P., and Stanley, J. C. "Sex Differences in Mathematical Ability: Fact or Artifact?" *Science* (December 12, 1980): 1262–64.

*Fennema, Becker, Wolleat, and Pedro, 1979. "Multiplying Options and Subtracting Bias" (set of four videotapes and Facilitator's Guide) is available at cost ($150.00) from Instructional Media Distribution Center, 1025 West Johnson St., Madison, WI 53706.

4. Burton, G. M. "Regardless of Sex." *Mathematics Teacher* 72, no. 4 (1979): 261–70.

5. Casserly, P. L. "Factors Leading to Success: Present and Future." In *Perspectives on Women and Mathematics,* edited by J. E. Jacobs. Columbus, Ohio: ERIC Clearinghouse for Science, Mathematics, and Environmental Education, 1978.

6. ———. "Factors Affecting Female Participation in Advanced Placement Programs in Mathematics, Chemistry, and Physics." In *Women and the Mathematical Mystique,* edited by L. H. Fox. Baltimore: Johns Hopkins University Press, 1980.

7. College Entrance Examination Board. *College-Bound Seniors: National Report, 1971–1979.* Princeton, N.J.: Educational Testing Service, 1979.

8. Equals. *Startling Statements.* Berkeley, Calif.: Lawrence Hall of Science, University of California, 1979.

9. Fennema, E. "Mathematics Learning and the Sexes: A Review." *Journal for Research in Mathematics Education* 5, no. 3 (1974): 126–39.

10. ———. "Sex-Related Differences in Mathematics Achievement: Where and Why?" In *Perspectives on Women and Mathematics,* edited by J. E. Jacobs. Columbus, Ohio: ERIC Clearinghouse for Science, Mathematics, and Environmental Education, 1978.

11. ———. "The Middle School: A Crucial Time for Females' Mathematics Learning." *National Council of Teachers of Mathematics 1982 Yearbook,* forthcoming.

12. ———, Becker, A.; Wolleat, P.; and Pedro, J. *Multiplying Options and Subtracting Bias.* Washington, D.C.: U.S. Department of Health, Education, and Welfare, 1979.

13. ———, and Carpenter, T. "Sex-Related Differences in Mathematics: Results from National Assessment." *Mathematics Teacher* 74, no. 7 (1981): 554–59.

14. ———, and Sherman, J. "Fennema-Sherman Mathematics Attitude Scales." Journal Supplement Abstract Series (JSAS) *Catalog of Selected Documents in Psychology* 6, no. 1 (1976): 31. (Ms. no. 1225)

15. ———, and ———. "Sex-Related Differences in Mathematics Achievement and Related Factors: A Further Study." *Journal for Research in Mathematics Education* 9, no. 3 (1978): 189–203.

16. Fox, L. H. "The Effects of Sex-Role Socialization on Mathematics Participation and Achievement." In *Women and Mathematics: Research Perspectives for Change.* National Institute of Education Papers in Education and Work: No. 8. Washington, D.C.: U.S. Government Printing Office, 1977.

17. National Council of Teachers of Mathematics. *The Mathematics Education of Girls and Young Women.* Reston, Va.: The Council, 1980.

18. Osen, L. *Women in Mathematics.* Cambridge, Mass.: Massachusetts Institute of Technology Press, 1974.

19. Sells, L. W. "The Mathematics Filter and the Education of Women and Minorities." In *Women and the Mathematical Mystique,* edited by L. H. Fox. Baltimore: Johns Hopkins University Press, 1980.

20. Vetter, B. M. *Opportunities in Science and Engineering.* Washington, D.C.: Scientific Manpower Commission, 1980.

ELEVEN

The Case for a New High School Mathematics Curriculum
Shirley Hill

The weary mathematics teacher of long experience may be pardoned a reaction such as "Here we go again. It must be time for another push to change high school mathematics." In retrospect, it does seem that periodically the public becomes acutely conscious of the reliance of our civilization upon technology and the dependence of technological development on a solid foundation in mathematics and science. This consciousness is just one obvious step removed from a concern for the quality of mathematics and science education as the bedrock of that foundation.

In 1957, when the Soviet Union launched Sputnik into space, there was an immediate hue and cry. The result was a considerable effort, with both public and private funding, to ensure the competitive position of the United States in technology by curriculum reform and teacher education in science and mathematics. Twenty years later a *Time* magazine feature article concluded that "twenty years after Sputnik" we had been successful. The evidence for this conclusion was the technological dominance of the United States.

In the meantime, the public's educational priorities had shifted dramatically. By the mid-seventies the concern was for issues of equity, for assuring that everyone achieved some minimal skill level. The largely unexamined assumptions of the back-to-basics movement proved attractive to much of the public and to many teachers.

THE REIGN OF THE TEST SCORE

The same period was the era of a remarkably unquestioned faith in test scores. For many people, including too many school administrators, successful education has become synonymous with high test scores. This has happened despite the fact that many who are critical of teachers and schools and school districts because of their students' performance on standardized tests do not know what those tests are testing or whether they even match the schools' stated objectives. This is *supreme faith* in the testing industry to interpret our educational priorities. In a time of rapidly changing needs it is a dangerous faith, because tests are a very conservative element in the curriculum dynamic.

Declines in test scores have been much publicized. Anyone paying attention knows of a decade-and-a-half decline in scores on the Scholastic Aptitude Test. It does not matter that the Educational Testing Service has tried to make clear that the SAT has a sole and narrow purpose, the prediction of performance in the first college year, and that it is not designed to provide evidence for evaluation of school programs. These technicalities escape the attention of the public and the media as references are made again and again to the SAT results as evidence of a failure of the nation's schools.

In 1977, an Advisory Panel on the SAT Score Decline conducted an exhaustive study of possible causes and reported in *On Further Examination* (1) that the causal factors fall into two quite different categories:

1. Changes in the SAT-taking population—a broadening that accounted for from two-thirds to three-fourths of the decline between 1963 and 1970, and one-quarter since 1970.
2. Changes in the practices of schools and the American social fabric (for example, more elective courses and fewer required courses, automatic promotions, grade inflation, absenteeism, reduction of homework, lowering of standards, lowering of college entrance standards, television, disintegration of family support, disruption in the life of the country, diminution of motivation for learning).

In short, there is a complex web of causes—both school and nonschool.

A NATIONAL CRISIS

Test scores alone have not generated the recent consensus of crisis in mathematics and science preparation. As with Sputnik, these alarms are related to perceptions of national needs and policy. The nation has serious concerns about its present and future ability to increase productivity, to fill trained personnel needs in industry and the military, and to maintain technological competitiveness. Many people recognize the critical role of mathematical skill and knowledge as a base for the technical training most of these needs demand.

The doubts have been aggravated of late by reports of the depth and extent of the mathematics and science components of the school programs in the Soviet Union, Japan, and West Germany. Concerns reached the highest levels of government and in 1980, President Carter requested an analysis by the National Science Foundation and the Department of Education. This analysis and the recommendations it generated are contained in the report, *Science and Engineering Education for the 1980's and Beyond* (5).

In an appendix, the report directly considers our competitive position:

> This concern centers on the important question of whether the United States faces a reduced ability, relative to other countries, to generate and incorporate technological change in its production and utilization of goods and services.

And noting anxiety, particularly about engineering and computer professions, it states:

> While the United States has current shortages and future shortfalls in these areas, the Soviet Union, Germany, and Japan are producing much larger proportions of engineers and applied scientists than we are. At the same time, these countries are educating a substantial majority of their secondary school population to a point of considerable scientific and mathematical literacy, in part because they apparently believe that such literacy is important to their relative international positions.

Further, the report leaves little doubt that its writers see a link between the fact that half our students opt out of mathematics and science study early in high school and these increasing shortages in the nation's pool of technically and technologically trained personnel.

WHAT ELSE IS NEW?

Let us return to our skeptical mathematics teacher who, having been around awhile, is thinking, "This sounds all too familiar. We've heard about crises before. Is it really any different now?" The fable about the boy who cried "Wolf!" comes at once to mind, but with the recollection that in the fable the wolf finally did come.

What is new, what is different now that lends particular urgency to a need to reexamine the high school mathematics program? There are, I believe, some new elements, as well as the acceleration of the pace of change. Mathematics and its applications have always had a mutually vitalizing relationship. The uses of mathematics have increased and mathematical models have found uses in areas never before touched by mathematical methods. Mathematical methods are pervasive and, in fact, have become our civilization's major problem-solving tool.

At the same time, mathematics itself has expanded its boundaries and changed some of its methods in profound ways. The term now used —"the mathematical sciences"—not just "mathematics"—conveys the importance not only of the discipline itself but of the delivery of its knowledge and techniques to the solution of a wider variety of problems.

Because of the pervasiveness of mathematics, the ordinary citizen, as well as the career user, needs a higher level of mathematical literacy to function effectively in an increasingly technological world.

The traditional program reflects an age when arithmetic skill and a little algebra (geometry was fine to "teach one to reason") were enough for most people; anything more was for engineers, scientists, or mathematicians. I am not certain that such beliefs were ever valid, but if they were, they certainly are valid no longer.

The NSF–Department of Education report puts it clearly:

> The contribution of science and technology to our security and prosperity rests on two bases. The first of these is the competence and inventiveness of the practitioners, the scientists, and engineers who design and build the system. But the second base is equally important to our overall success as a Nation. This second base consists of the overwhelming portion of our population which has no direct involvement in science and technology, or with the science and engineering community. They are indirectly involved through their influence on the governmental and industrial sources of funding for scientific and technological endeavors. They are involved in the regulatory and policy decisions that set directions for scientific inquiry and technological development. They reap the benefits of science and technology. Many need some knowledge of science and technology to do their jobs well. However, the current trend toward virtual scientific and technological illiteracy, unless reversed, means that important national decisions involving science and technology will be made increasingly on the basis of ignorance and misunderstanding. (5)

But the new element that promises change that can truly be called "revolutionary" is the computer and its growing availability. Education at all levels must respond immediately and forcefully to the computer and its role. At present, educational programs are not nearly keeping pace with the needs engendered by computer technology and usage.

SOME RECOMMENDATIONS

In 1978, the Mathematical Association of America organized a conference on Prospects in Mathematics Education in the 1980's (PRIME–80). The conference culminated in agreement on a number of recommendations (2). Some relate directly to high school mathematics, some relate to collegiate mathematics but have implications for high school preparatory programs. Among the latter are the following:

The MAA should undertake to describe and make recommendations on an alternative to the traditional algebra-calculus sequence as the starting point for college mathematics.

Many students will take only a few mathematics courses in college and will benefit most from courses at the freshman-sophomore level that include ideas, methods, and applications from statistics, probability, computing science, and applications to real-world situations through model-building methods.

Every college graduate should have some minimal knowledge of the mathematical sciences.

Nowadays people constantly encounter arguments based on numerical data, assertions stated in terms of probability, and situations that involve applications and uses of computing machines and algorithms.

Noteworthy in these statements are the references to the newer mathematical sciences—statistics and probability, computing. The traditional high school curriculum is largely conditioned by the absolute centrality of calculus in all college mathematics—a dominance brought into question by the MAA statements.

This question suggests the need to look closely at the high school program to see if it does or should reflect the stress on probability, statistics, and computing. Is a singular college preparatory track sufficient now? Perhaps the potential users in business, management, and social sciences could benefit from a somewhat different program from that of the future engineer, chemist, or physicist.

An MAA recommendation bearing directly on the secondary curriculum hints at this possibility:

The MAA should be alert to, and inform other appropriate agencies of, the possibility that secondary school mathematics programs might get out of step with developing college programs aimed at subsequent careers in computing, statistics, and other areas of applied mathematics.

WHAT SHOULD BE REQUIRED?

There is increasing evidence around the country of a belief that present mathematics requirements (typically one year in grades 9–12) are inadequate. In a number of states and local districts there is pressure from boards, state agencies, and teacher groups to raise the requirements.

There appears to be popular support for such change. In a Gallup poll asking which subjects are essential for all high school students, mathematics ranked highest (97 percent). More informative, however, are the data from a 1979 survey by the National Council of Teachers of Mathematics, with NSF funding. The project, entitled *Priorities in School Mathematics* (PRISM), surveyed a broad range of populations—professional and lay (4).

One question asked the lay sample how many years of high school (grades 9–12) mathematics should be required. For the college-bound student, almost half would require four years and 83 percent at least three years. For all students: four years (15 percent), three years (25 percent), two years (47 percent), or a total of 87 percent supported at least two years.

Particular concern was expressed in the PRIME–80 Conference of the MAA about high school preparation in mathematics. It was recognized that a part of the problem is the failure of half the student population to elect mathematics study beyond tenth grade. A recommendation that ensued was "to encourage all high school students to take at least three years of high school mathematics."

But to "encourage" is not to "require." Who is to do the encouraging? Presumably, teachers and counselors, and perhaps parents. The data indicate, however, that such encouragement has not been forthcoming; nor has it been effective.

It might be expected that the lay public would be more enthusiastic about raising requirements than would the professionals, who correctly foresee the enormous difficulties involved. But a realistic assessment of these difficulties does not obviate the need of both individual and society for a higher level of mathematical literacy. It should not be an excuse for papering over the problem.

A professional organization of mathematics teachers, the National Council of Teachers of Mathematics, confronted this issue head-on in one of the recommendations in its *Agenda for Action: Recommendations for School Mathematics of the 1980's,* released in 1980 (3).

The recommendation states:

> *More mathematics study must be required for all students and a flexible range of options should be designed to accommodate the diverse needs of the student population.*

And it specifically adds:

> *At least three years of mathematics should be required in grades 9 through 12.*

This recommendation often leaves school personnel alternately laughing or gasping and stammering. The immediate objections such as "That won't be the best program for everyone." or "Some students can't get through that level of mathematics." proceed from an erroneous assumption. They assume that what is meant is keeping all students longer in the same programs, those now existing. In other words, they imagine everyone being herded through algebra–geometry–algebra. This might indeed be a disaster.

The NCTM recommendations have something else in mind. They no sooner recommend three years for all students, than they issue this caveat:

> To say that most students should study more mathematics is not to say that it should be the same mathematics for all . . . In fact, such a recommendation poses a tremendous challenge to curriculum developers and school districts to devise a more flexible range of options, a diversified program to meet a variety of interests, abilities, and goals. (3)

The rationale for the general recommendation refers to at least four different clienteles for high school mathematics. It suggests "more than a single college-preparatory program," building on the growing need in the social sciences, management science, business for a foundation in statistics and probability, mathematical models and computer science, as well as retaining the traditional precalculus preparation.

It speaks of the expanded needs for mathematics skill among those whose postsecondary educational experience will be in highly skilled technical areas or in vocational skill training.

Finally, it speaks to the future citizen and consumer—regardless of career or job:

> For those whose formal education will end with high school, the needs of citizen and consumer for increasing mathematical sophis-

tication dictate a collection of courses based on consumer and career needs, computer literacy, and quantitative literacy. (3)

A few other points from the NCTM recommendations deserve study:

- Algebra should be included in the programs of all capable students to keep their options open.
- For many students, algebra should be delayed until a level of maturity and basic mathematical understanding permit their taking full advantage of a significant algebra course. Significant mathematics courses should be available to these students early, not just the traditional general mathematics review or prealgebra course.
- Consumer mathematics should develop a broader quantitative literacy and should consist primarily of work in informal statistics.
- All high school students should have work in computer literacy, the hands-on use of computers, and the applications of computers.
- All students who plan to continue their study of mathematics beyond high school or to use it extensively in technical work or training should be enrolled in mathematics courses throughout their last high school year. (3)

On the urgency of the need to change things, both in mathematics requirements and in greater diversity of course offerings, I rest my case. I have no illusions about the difficulties—difficulties aggravated by financial woes of schools, smaller enrollments and school closings, class sizes, problems of student motivation, lack of public support, shortage of teachers certified in mathematics. But the status quo is simply indefensible.

What is needed are some new courses, some imaginative revamping of existing courses, some creative and effective integration of computers into most courses, good new textbooks and other materials, good computer software. Ongoing teacher support systems, perhaps on the order of the old NSF institutes and academic-year institutes, could also be used.

The implications of the case I have outlined suggest some starting points for curriculum rebuilding. For the traditional mathematics user, the traditional college preparatory program should, for the most part, remain as is. Changes would be the incorporation of computers and computer methods, and a greater stress on applications and modeling, in most courses.

Strongly recommended, however, is an alternative path for other fully capable students who will also become mathematics users in areas where the mathematics orientation is more recent. Such a program would stress statistics and computer science rather than calculus.

In terms of national humanpower needs, there is probably no category so vital as the potential technician—the highly skilled worker in industry, government, and the military. The high school curriculum builders have rarely considered this student. He or she may pursue postsecondary training in technical institutes or on the job in industry or the military. As technology becomes more complex, the skills needed by the technician require a strong mathematics background. This preparation needs special attention as well as the cooperation of teachers, curriculum builders, and industry to determine the best way to meet these demands in the high school program. As a beginning, some components of the existing curriculum, computer experience, and perhaps a new course in technical applications could provide the basis.

For the future citizen whose career use of mathematics will be minimal, a course in computer literacy and a course in strong consumer applications dealing largely with informal statistics skills and knowledge are recommended. For those who are capable and those who may attend college, algebra and geometry should be added. For those few who are not capable of completing algebra, a senior year course in personal finance management could complete a three-year program.

Ideally, a computer literacy course with some hands-on computer experience would be given at eighth grade. Then subsequent courses could build on this foundation by integrating computers as tools. For some students, a substantial computer science course could come late in high school.

The need for a sound but very elementary course in informal statistics—finding, organizing, and using information; drawing inferences from data; conducting experiments—should be stressed. Such a course for consumers would be beneficial to several categories of student.

Only a few *new* courses need to be designed. Existing courses should always be revised and updated in any case, and computers *must* be incorporated.

The difficult part will be to cope with the logistics of planning individual programs that truly accommodate the wide diversity of needs, interests, and capabilities, yet provide a sound and useful background for all students. Two things should be given top priority in such curricular design. The first is flexibility. As much as possible, students should be able to move to a different program, to move laterally without becoming trapped in a tracking system. Without a crystal ball, the precise needs of each individual cannot be predicted with confidence.

This suggests the other priority—to keep options open as long as possible, to keep broad career choice viable. This explains the NCTM recommendation to include algebra if at all possible, even if the course is deferred to later grades. Students should avoid closing doors on future choice.

Such curriculum change is a formidable task, but it is not an impossible one. The future will not wait and planning should begin now.

REFERENCES

1. Advisory Panel on the Scholastic Aptitude Test Score Decline. *On Further Examination.* New York: College Entrance Examination Board, 1977.

2. Mathematical Association of America. *Proceedings of a Conference on Prospects in Mathematics Education in the 1980's.* Washington, D.C.: The Association, 1978.

3. National Council of Teachers of Mathematics. *An Agenda for Action: Recommendations for School Mathematics of the 1980's.* Reston, Va.: The Council, 1980.

4. ———. *Priorities in School Mathematics.* Reston, Va.: The Council, 1981.

5. National Science Foundation–Department of Education. *Science and Engineering Education for the 1980's and Beyond.* Washington, D.C.: U.S. Government Printing Office, 1980.

The Contributors

Barbara J. Bestgen is an Elementary Mathematics Specialist in the Parkway School District, St. Louis. A former researcher and high school mathematics teacher, Ms. Bestgen is a coauthor of *Keystrokes: Calculator Activities for Young Students—Addition and Subtraction* and *Keystrokes: Calculator Activities for Young Students—Multiplication and Division,* and a frequent contributor to professional journals.

Thomas P. Carpenter is Professor of Curriculum and Instruction at the University of Wisconsin–Madison. He directed a National Science Foundation project to interpret the results of the second mathematics assessment of the National Assessment of Educational Progress.

Mary Kay Corbitt is an Assistant Professor in the Departments of Mathematics and Curriculum and Instruction at the University of Kansas, Lawrence, and a consultant to the National Assessment of Educational Progress. A former high school mathematics teacher, Dr. Corbitt's wide range of professional activities has included project evaluation and preparing and editing mathematical reports. She is a frequent contributor to mathematics journals.

Elizabeth Fennema is a Professor in the Department of Curriculum and Instruction at the University of Wisconsin–Madison. She has taught at all educational levels and has conducted extensive research on sex-related differences in education.

James T. Fey is an Associate Professor at the University of Maryland. He is a former high school mathematics teacher, research associate for the Secondary School Mathematics Curriculum Improvement Study, consultant to UNESCO projects in developing countries, and Executive Secretary of the National Advisory Committee on Mathematics Education. A contributor to numerous professional and research journals, Dr. Fey is the author of *Long, Short, High, Low, Thin, Wide.*

J. D. Gawronski is Director of Planning, Research, and Evaluation for the San Diego County Department of Education, and a consultant to school boards and districts on computer education and computer literacy. Dr. Gawronski is the author of several articles on the use of computers in the classroom.

Shirley Hill is a Professor of Education and Mathematics at the University of Missouri–Kansas City, and a trustee of the Conference Board of the Mathematical Sciences. Dr. Hill is a past president of the National Council of Teachers of Mathematics and a former chair of the U.S. Commission on Mathematical Instruction. She is a coauthor of *Overview and Analysis of School Mathematics, K–12,* and the author of books on geometry and logic, and of numerous articles in professional journals.

Mary Grace Kantowski is an Associate Professor of Mathematics Education at the University of Florida. A former secondary school mathematics teacher, Dr. Kantowski's experience includes training teachers at the elementary and secondary levels; serving as consultant to several projects on problem solving; and directing projects on research, development, and training related to problem solving and the microcomputer sponsored by the National Science Foundation and the National Institute of Education.

Henry S. Kepner, Jr., is an Associate Professor of Mathematics Education at the University of Wisconsin–Milwaukee. Dr. Kepner is a past president of the Wisconsin Mathematics Council, a former chair of the Instructional Affairs Committee of the National Council of Teachers of Mathematics, and a former mathematics teacher in junior and senior high school. He has also taught computer science and techniques to both elementary and secondary students, and has published widely in mathematics education journals. He is coauthor of *Reading in the Mathematics Classroom* and editor of *Computers in the Classroom* published by NEA.

Mary Schatz Koehler is a doctoral student in mathematics education and a project assistant at the University of Wisconsin–Madison. Ms. Koehler is a former high school mathematics teacher and mathematics department chairperson, and is a contributor to professional journals.

Mary Montgomery Lindquist is Chair of the Mathematics/Science Department of the National College of Education, Evanston, Illinois. She is also Chair of the Research Advisory Committee of the National Council of Teachers of Mathematics, a board member of the *Journal for Research in Mathematics Education,* and a member of the team sponsored by NCTM and funded by the National Science Foundation to interpret the results of the second mathematics assessment conducted by the National Assessment of Educational Progress.

Terry E. Parks is Executive Director of Program Development and K–12 Director of Mathematics/Computers for Instruction in the Shawnee Mission, Kansas, Public Schools. A former science and mathematics teacher at the junior and senior high school levels, Dr. Parks also serves as southwestern regional representative to the Regional Services Committee of the National Council of Teachers of Mathematics.

Robert E. Reys is a Professor of Mathematics Education at the University of Missouri–Columbia. Dr. Reys has served as president of the Missouri Council of Teachers of Mathematics, as general editor of the yearbooks of the National Council of Teachers of Mathematics from 1976 to 1980, and as a mathematics consultant. A frequent contributor to professional journals, he is the author of *Methods of Teaching Elementary School Mathematics* and *Algebra for Elementary Teachers,* and the coauthor or editor of several other publications.

Sherilyn Seitz is a fourth grade teacher at Highlands Elementary School, Shawnee Mission, Kansas. She has also taught at the junior high school level and has had experience as a senior high school guidance counselor. Ms. Seitz serves as Elementary Director of the Kansas City Area Teachers of Mathematics.

Gwen Shufelt is a Lecturer in Mathematics and Mathematics Education at the University of Missouri–Kansas City, and editor of the 1983 Yearbook of the National Council of Teachers of Mathematics. She is a past president of the Georgia Council of Teachers of Mathematics and a former mathematics coordinator and high school mathematics teacher. Ms. Shufelt also served on the Board of Directors of the NCTM, on the NCTM task force that prepared *Recommendations for School Mathematics of the 1980's,* and on the Program Committee for the Fourth International Congress on Mathematics Education.

Albert P. Shulte is Director of Mathematics Education, Oakland Schools, Pontiac, Michigan, and editor of the 1981 Yearbook of the National Council of Teachers of Mathematics, *Teaching Statistics and Probability.* A past president of the Michigan Council of Teachers of Mathematics and of the Detroit Area Council of Teachers of Mathematics, he served as a member of the American Statistical Association–NCTM Joint Committee on the Curriculum in Statistics and Probability. He is the coauthor of two textbooks and several supplementary publications.

Marilyn N. Suydam is an Adjunct Professor of Mathematics Education at The Ohio State University, Columbus; Associate Director of the ERIC Clearinghouse for Science, Mathematics, and Environmental Education; and Director of the Calculator Information Center. A former elementary school teacher, she served as editor of *Developing Computational Skills,* the 1978 Yearbook of the National Council of Teachers of Mathematics.

AN
12

A